The Science of Science Communication II

Summary of a Colloquium

Held on September 23-25, 2013, at the National Academy of Sciences in Washington, D.C.

NATIONAL ACADEMY OF SCIENCES
THE NATIONAL ACADEMIES

THE NATIONAL ACADEMIES PRESS
Washington, D.C.
www.nap.edu

THE NATIONAL ACADEMIES PRESS **500 Fifth Street, NW** **Washington, DC 20001**

This volume is based on the Arthur M. Sackler Colloquium of the National Academy of Sciences, "The Science of Science Communication II," held September 23-25, 2013, at the National Academy of Sciences building on Constitution Avenue in Washington, DC.

International Standard Book Number -13: 978-0-309-29200-9
International Standard Book Number -10: 0-309-29200-X

Additional copies of this book are available from the National Academies Press, 500 Fifth St., NW, Keck 360, Washington, DC 20001; (800) 624-6242 or (202) 334-3313; http://www.nap.edu.

Cover art: In this 2013 commissioned work by Heather Larkin, the landscape of science communication appears as a patchwork of regions—or conceptual contexts—each characterized by unique features, dimensions, and hues. Communication bridges the regions in complex combinations of messages between individuals, groups, and institutions.

Printed in the United States of America

Contents

Preface

Successful scientists must be effective communicators *within* their professions. Without those skills, they could not write papers and funding proposals, give talks and field questions, or teach classes and mentor students. However, communicating with audiences outside their profession—people who may not share scientists' interests, technical background, cultural assumptions, and modes of expression—presents different challenges and requires additional skills. Communication about science in political or social settings differs from discourse within a scientific discipline. Not only are scientists just one of many stakeholders vying for access to the public agenda, but the political debates surrounding science and its applications may sometimes confront scientists with unfamiliar and uncomfortable discussions involving religious values, partisan interests, and even the trustworthiness of science.

In response to these problems, the National Academy of Sciences has hosted two Sackler colloquia on The Science of Science Communication. These events brought together leading social, behavioral, and decision scientists to familiarize one another, other scientists, and communication practitioners with current research that can improve the communication of science to lay audiences. In the Sackler colloquia tradition, the meetings also allowed social and natural scientists to identify new opportunities for collaboration and advancing their own research, while improving public engagement with science.

The first colloquium, and accompanying special issue of the *Proceedings of the National Academy of Sciences*,[1] included research in science

[1] See http://www.pnas.org/content/110/Supplement_3.

education, communication, medicine, and decision science. The second colloquium, reported here, incorporated social and cognitive psychology, political science, mass communication, cultural anthropology, business, and social network analysis. It, too, will be captured in a special issue of *PNAS*, with articles summarizing relevant research areas in those sciences and applying them to domains as diverse as climate change, nanotechnology, and medicine.

Following 2 days of talks and panels, the second colloquium hosted day-long work sessions, bringing together subject-matter experts, researchers in the sciences of communication, and communication professionals. Together they developed communication strategies for four topics: climate change, nanotechnology, obesity and nutrition, and evolution. More than 500 people attended the colloquium, while over 10,000 joined concurrent webcasts or visited the archived recordings.[2]

As reported here, speakers encouraged scientists to approach communication in new ways and to create the infrastructure needed to link scientists, communication professionals, and their audiences. Speakers provided evidence-based guidance on how to listen to others so as to identify their information needs, ways of thinking about the world, and the cultural stereotypes regarding scientists. They delved deeply into the incentive systems that shape what scientists study and how they report their work, the subtle changes in "framing" that can influence how messages are interpreted, the complex channels that determine how messages flow, and the potential politicization of scientific evidence.

Speakers were also challenged to go beyond their disciplinary pursuits in order to understand the problems that face those scientists who attempt to communicate their work and collaborate with scientists from other disciplines. In the spirit of the Sackler colloquia, those collaborations can lead to research that would not have occurred without working together in a common cause. As a result, the enterprise can be as beneficial for the sciences of communication as for the communication of science.

Finally, the colloquium organizers would like to thank the National Science Foundation, the Gordon and Betty Moore Foundation, *Science*, and COMPASS for their valuable financial support of the colloquium summarized in these pages.

Ralph J. Cicerone
Baruch Fischhoff
Alan I. Leshner
Barbara A. Schaal
Dietram A. Scheufele
Co-organizers

---

[2] See http://bit.ly/sackler.

1

The Sciences of Communication

Several years ago, a committee appointed by the National Academy of Sciences began working on the third edition of the book *Science and Creationism.* The first two editions of the book had been widely cited in legal cases, said Barbara Kline Pope, executive director for communications at the National Academies and director of the National Academies Press, in her introductory remarks. But the committee and its staff wanted the third edition to have a bigger impact. At the time, intelligent design creationism was a relatively new concept, and it was being pushed into science classrooms by vocal and well-financed groups. The committee decided that formal audience research was warranted—despite the time and cost it would add to the project—to gauge what people believe about evolution, intelligent design, and creationism.

Just as the facilitators' guide for the focus groups was being written, Judge John Jones issued his decision in the case *Kitzmiller v. Dover Area School District*, ruling that intelligent design "cannot uncouple itself from its creationist, and thus religious, antecedents." The staff working on the book thought that this message would be a "slam dunk," said Pope. They felt that the message would resonate strongly with audiences and directed the focus group facilitators to point out that a judge had decided that intelligent design is a form of creationism and thus religion and that teaching it in science classrooms is therefore unconstitutional and illegal.

Pope was watching the live focus groups on her computer while eating dinner. "I slowly lowered my fork to my plate, and my jaw dropped with it. I saw backs seizing up and eyes getting squinty, and one guy said,

'No judge is going to tell me how to run our schools.' I saw the rest of the participants clenching their jaws and nodding enthusiastically."

The staff was shocked at how wrong they had been, especially when a quantitative survey uncovered the same attitudes. Relying on their intuition about effective messages would have been "a very bad idea," said Pope, and the third edition of *Science, Evolution, and Creationism* (NAS/IOM, 2008) is a very different document than it would have been had audience research not been done.

The Science of Science Communication II colloquium was similarly devoted to using the best available evidence to guide science communication. The colloquium was built on the first Science of Science Communication colloquium,[1] but it sought to dig deeper into the methodologies, analyses, and findings of science communication research. It also featured, on the third day of the colloquium, concurrent workshops on four pressing topics—evolution, climate change, nanotechnology, and nutrition and obesity—where researchers and practitioners could develop research-based insights on communication strategies that would have immediate application.

LAY NARRATIVES AND EPISTEMOLOGIES

Science communication occurs through artifacts, including language, diagrams, and other representations. These artifacts both reflect the cultural assumptions of their creators and reinforce different ways of seeing the world, said Douglas Medin, the Louis W. Menk Professor of Psychology at Northwestern University. Science communication, therefore, needs to pay attention both to the artifacts with which it is conducted and to the different ways people have of looking at the world.

As an example of cultural differences in perspectives, Medin cited cognitive research comparing East Asians, typically Chinese, Japanese, and Koreans, with westerners, typically people from the United States. East Asians tend to pay more attention to background information, while westerners attend more to focal objects. For example, when shown successive pictures that look very similar, East Asians are much better at detecting background changes, while westerners are better at detecting foreground changes. Another study found that western paintings have three to four times as much representation devoted to faces, while East Asian portraits include more background information. The same difference was reflected in the aesthetic preferences of East Asians and westerners.

---

[1] See http://www.nasonline.org/programs/sackler-colloquia/completed_colloquia/science-communication.html.

Native Versus European American Perspectives

Medin and his colleagues at Northwestern have been involved in a collaborative research partnership with the American Indian Center of Chicago and the Menominee Nation of Wisconsin. Four thousand to five thousand Menominee, who are the oldest continuous residents of Wisconsin and are well known for their sustainable forestry practices, live on tribal lands in and around three small communities. Interviews with Menominee and European American parents and grandparents revealed large differences in distancing discourse. When European American parents and grandparents were asked about the five things they would like their children or grandchildren to learn or know about the biological world, they talked about nature as an externality. They wanted their children to respect nature and know they have a responsibility to take care of it. Native American parents and grandparents were much more likely to say that they wanted their children to understand that they are a part of nature.

Another example of distancing discourse comes from depictions of ecosystems in publications by westerners. Virtually none include humans as part of the representation, suggesting that westerners generally think of themselves as outside of ecosystems.

Another demonstration of differing perspectives comes from an analysis of children's books written by Native American and European American authors. The illustrations by Native Americans tend to have closer, more personal, and more wide-angle representations. As a result, they provide more alternative perspectives. The books by European American authors were more likely to have straight-ahead perspectives at eye level. The Native American books were more likely to provide the perspective of an actor in the scene by using an over-the-shoulder or embodied representation.

The texts of the books also differed. Native American–authored books were more likely to mention seasonal cycles, native animals, and objects that, in a western perspective, would be part of the background.

Conceptions of Nature

These results parallel those from cognitive experiments on conceptions of nature. For example, when Native American and European American adults in rural Wisconsin were asked to describe the last time they went fishing, the median point at which European Americans used the word "fish" was the 27th word, whereas the median for the Native American Menominees was the 83rd word. The Native Americans were much more likely to supply context and background information—so much so that some never mentioned fish at all.

In another study, about 40 percent of Menominee children spontaneously imitated or took the perspective of the animals when shown pictures containing those animals, while no European American children did so.

In a set of related studies, fishing experts were asked to sort 44 local species of fish into categories that made sense to them. European American experts tended to sort the species either in terms of taxonomic relations, such as the bass family, or by goals, such as large and prestigious game fish. Menominee experts were much more likely to sort ecologically or by habitat, such as fish found in fast-moving water. However, when the groups were asked to sort the fish by habitat, the differences disappeared, indicating that the intergroup differences involve the organization of knowledge rather than knowledge per se. In a follow-up study, Menominee experts were more likely to recognize positive reciprocal relationships among fish species, such as the reciprocal eating of spawn, fry, and small fish, while European American experts mentioned fewer relations, and those that they did mention primarily involved adult fish.

Implications for Science Communication

These findings have some important implications for science communication, Medin concluded. Distancing and outsider perspectives can undermine engagement with science. The use by researchers of terms such as "the public" also can be distancing and homogenizing.

In addition, community members bring many skills to an exchange with experts. In work the collaboration has done on community-based citizen science, participants have had strong backgrounds in chemistry, hydrology, and forestry.

Finally, mismatches between lay epistemologies and orientations implicit in communication may be a source of alienation, Medin said.

Artifacts That Shape Perspectives

The research described by Medin raises intriguing questions about the everyday artifacts that shape views of humans in nature, said Ann Bostrom, the Weyerhaeuser Endowed Professor of Environmental Policy at the University of Washington. For example, how are children's perspectives shaped by what they see? And how readily can these perspectives be changed?

Bostrom addressed the first question by considering several video games popular among children. In the game Knytt Underground by Nicklas "Nifflas" Nygren, children explore a natural landscape from a third-person perspective. In the game Minecraft, on the other hand, children play from a first-person perspective. Some games, such as Kerbal

Space Program, can be played from either a first-person or third-person perspective, though one perspective may be easier to play than the other, depending on the game.

Other visual representations also can be either first or third person. For example, popular earthquake films can leave the viewer outside the scene by taking a third-person perspective or draw a viewer into the scene with a first-person perspective.

The basic metaphor for these representations, as Stanford psychologist Barbara Tversky has pointed out, is proximity, and proximity has an influence on cognitive progresses. Because humans have embodied minds and points of view, our spatial orientation affects the speed with which we process information. For example, from top to bottom is easier for us to discriminate than from side to side, because our bodily axes and asymmetry affect how we process information.

With regard to artifacts, readers form mental images from texts. These images can have a variety of perspectives due largely to the abstractness of texts. Translations offer a good example. One such translation is of an ancient poem by Wang Wei entitled "Deer Park":

> There seems to be no one on the empty mountain…
> And yet I think I hear a voice,
> Where sunlight, entering a grove,
> Shines back to me from the green moss.

A more recent translation is by the poet Gary Snyder:

> Empty mountains: no one to be seen.
> Yet—hear—human sounds and echoes.
> Returning sunlight enters the dark woods;
> Again shining on the green moss, above.

Changes in perspectives also characterize mental models of hazardous processes, Bostrom observed. For example, about 10 percent of people asked in one study about climate change said that they have direct experience with climate change—for instance, through changes in the seasons. In other cases, people recruit mental models that do not depend on personal experience.

Stories create meaning, Bostrom concluded. They make causality concrete and close. But can stories alone enable readers to shift fluently between points of view? The broad environmental expertise of the Menominee may enable them to shift fluently between perspectives, as well as to better distinguish between correlation and causation. But the question of which artifacts matter for science communication remains largely unanswered.

Multiple Representations of Knowledge

Multiple representations of knowledge are also an important consideration in science education, said Kevin Dunbar, professor of human development and quantitative methodology at the University of Maryland in College Park. In the past, science education has been oriented toward filling students with facts that they can repeat back on tests. More recently, education has emphasized the construction of knowledge at a social as well as an individual level. By having students pose questions on nature, they are expected to learn through discussion. Students also learn by confronting their naïve conceptions with the results of classroom experiments.

However, what if multiple valid epistemologies exist, Dunbar asked. Similarly, if different cultures have different views of science, is the result different sciences? For example, do different Native American nations have different epistemologies, and if they do, should the differences be reconciled or should those differences be used to communicate science?

Dunbar has worked as a consultant on the Trail of Time in the Grand Canyon. At least 12 American Indian tribes think of the Grand Canyon as their homeland, and these tribes can have different epistemologies related to time. How can these differences be presented in positive way rather than one group being wrong and one being right? This is a major goal of science communication, said Dunbar—to deal with differences in a constructive way.

Finally, Dunbar mentioned ongoing research into how culture changes the brain. For example, are epigenetic changes a mechanism for the embodiment of cultural knowledge? Studies looking at epigenetic changes following educational interventions have been "very suggestive," but much more work needs to be done to know whether biology can inform science education.

Multiple Cultures in Science

A prominent topic of discussion during the question-and-answer period was the influence of multiple cultures on science. As Medin pointed out, the way that science gets done depends on the cultures of the people who are doing it, and multiple approaches to science make for strong science. This is a strong argument, he said, for diversity in research teams and in science education. For example, Medin's collaboration had a diverse research team to approach problems from multiple perspectives. However, attracting diverse groups to science can be difficult, Medin added. Unlike medicine, which often reflects the deepest values of medical students, the study of science often does not always allow students to express their deepest values.

As another example of the value of multiple perspectives, Medin cited research in primatology, which has made progress both from a western orientation that sees researchers as distant from nature and from an eastern perspective that sees researchers as part of nature. Dunbar pointed to the differences in culture between U.S. molecular biology laboratories, which tend to be inductive, and Italian laboratories, which are much more deductive, even though the laboratories produce papers that are very similar and are published in the same journals.

The presenters also discussed alternatives to the term "the public," given the diversity of audiences that science communicators would like to reach. Bostrom suggested using the names of professions, the groups with which people are involved, the communities of which they are members, the roles they play in relation to the use of science, or the technologies they use.

Discussion also revolved around the extent to which perceptions shaped by culture can be changed. Medin pointed out that cultural artifacts can have causal force. For example, presenting Americans with typical Japanese scenes can lead the American participants to become relatively better at detecting background changes, though perceptions are probably both chronic and flexible, he said. Bostrom observed that perceptions are reinforced by the roles people take in the world, which tend to reinforce both their cultural positions and their ways of thinking about the world. Yet psychology research shows that people's views often depend on context as well.

Medin pointed out that there are hundreds of federally recognized Native American tribes, and so results from one tribe cannot generalize to all, and great diversity exists even within a single culture. As Bostrom added, even with professional groups such as hurricane forecasters and emergency responders, differences in the mental models within a group are larger within the differences among groups.

MOTIVATED AUDIENCES: BELIEF AND ATTITUDE FORMATION ABOUT SCIENCE TOPICS

Audiences' motivations as human beings affect how they interpret science communications, observed Susan Fiske, Eugene Higgins Professor, Psychology and Public Affairs, at Princeton University. With climate change communications, for example, people who pay more attention to politics in the news have perceptions polarized away from those of people who pay more attention to scientific and environmental stories in the news. Furthermore, over time the amount of skepticism about climate change has increased, despite the increasing scientific consensus on the role of human beings in the changing climate. And to the extent that information about climate change is getting through, members of the public are getting more alarmed, which is not necessarily the message that policy makers or scientists wish to convey.

The first step in reconciling differences between scientific and public perspectives is to recognize that scientists are not the sole source of valid information, Fiske said. The second step is to recognize that both cognition and emotions, or affect, influence perceptions and interpretations. Together, cognition and emotions create motivation, and persuasion works best when both factors are taken into account.

People are not idiots, Fiske said. The public now knows more about climate change than in the past, and it generally can distinguish science from nonscience. Some have a classical view that science yields a single true picture of the world, while some have a more modern view that science can produce multiple answers that have to be negotiated and debated.

The Credibility of Communicators

To be credible, communicators need both expertise and trust, Fiske said. If someone is seen as an expert on a topic, other people tend either to agree with that person or at least think about a message. But communicators also need to be trusted to be effective, which means that they need to be seen as having a motivation to be truthful and accurate.

Trust has been a largely neglected topic in the science of science communication. In general, people trust those who they think are like themselves. People who belong to a group have a shared reality and a motivation to share understandings. "This is human nature," explained Fiske. "People trust people who they think share their values [and] goals. . . . This is a core insight within social and behavioral science." Group membership provides a sense of control over one's environment and circumstances. It also enhances feelings of self-worth. Thus, both cognitive and affective factors affect trust.

Warmth and Competence

Fiske and her colleagues have developed a framework for understanding the social and cultural landscape of groups. The first question people ask, in identifying whom to trust, is whether a person is friend or foe. If a person is seen as being on the same side or sharing the same values, they are seen as trustworthy and warm.

The second question people ask is whether the other can act on their own intentions. In other words, is the other person competent so that their acting on those intentions will produce a desired (or undesired) outcome?

The combination of warmth and competence produces a two-dimensional space that people use to interpret communications. For example, according to research by Fiske and colleagues that was under review at the time of the colloquium, polling data from more than 30 countries demonstrate that the middle class, white people, and blue-collar people are seen as high in both warmth and competence. Poor people and teens are seen as being neither warm nor competent—and the results for poor people are true all over the world, said Fiske. Children and old people are seen as being well intentioned but not very competent, while rich people are seen as competent but not warm, again all over the world.

Fiske noted that responses to people in the four quadrants of this two-dimensional graph fall into four emotional categories. Cold and incompetent people tend to be treated with disgust, warm and incompetent people with pity, competent and cold people with envy, and competent and warm people with pride.

According to pilot data collected online, when professionals are assessed against the dimensions of warmth and competence, researchers and scientists are seen as competent but cold, while professors and teachers are seen as both competent and warm—though not as competent and warm as doctors and nurses. When asked about the emotions felt toward these groups, researchers and scientists, in keeping with their position in the two-dimensional space, were more often the subjects of envy. People cooperate with envied groups because they have needed resources, including knowledge. But these groups can be attacked in times of instability, which creates a dangerous ambivalence. Envy implies that "you have things that I respect and I'd like to have, and I'd like to take them away from you," Fiske said.

Cold competence also can create resentment. For example, envied groups can be the object of Schadenfreude—the sense of pleasure at someone else's misfortune. When electrodes are connected to people's facial muscles, images of someone from an envied group getting his or her comeuppance often generate smiles. "When a guy in an Armani suit gets splashed by a taxicab or sits in gum on a park bench, people smile. They can't help it."

Increasing Warmth

People tend to believe that scientists and researchers are competent but do not trust their intentions. For example, when asked about the intentions of climate scientists, some answered that scientists might lie with statistics, complicate simple stories, feel themselves superior to non-scientists, pursue a liberal agenda, or provoke and hurt big corporations.

The most common answer is that scientists might slant research to get research funding. "That's our Achilles heel," said Fiske.

Public perceptions of science and scientists are more polarized today than they have been in the past. For example, even the proposal to name a science laureate for the United States has encountered political resistance. Opponents felt that a science laureate would have a pulpit to talk about values rather than science and disseminate a political agenda.

The news is not all bad, Fiske continued. Scientists gain a measure of trust because they are seen as interested in educating the public, preserving the environment, and saving humanity. People respect the educational mission of scientists and researchers in the same way they do those of teachers and professors. People also tend to trust an impartial agenda and not trust a persuasive agenda, which argues for separating science communications from the policy implications of those communications.

Scientists would be more trusted if they emphasized deliberation rather than persuasion, Fiske said. Scientists need to respect the intelligence of their audiences. They need to convey information and resist issuing policy conclusions unless they clearly label such conclusions as their own opinions.

One way to warm up scientists would be to emphasize their service to the public through forums such as the National Academy of Sciences, Fiske emphasized. The teaching role of scientists also generates positive reactions among members of the public. Similarly, letting people know why someone went into science, and having a diversity of people working in the sciences, can increase trust. By clearly expressing motivation, scientists can establish credibility and not be treated simply through stereotypes. Though the incentive structure in U.S. universities remains oriented toward research, teaching in universities, through its influence on the next generation of managers and policy makers, can improve the public's trust of the scientific community.

Influences on Perception

Craig Fox, Ho-Su Term Chair in Management at the UCLA Anderson School of Management, elaborated on the unconscious allure of in-group positions and the polarized political environment that tends to drive people's perceptions apart. In one experiment, people were asked what their political affiliation was, after which they were asked whether they wanted to invest in a conservative, moderately conservative, moderately risk-tolerant, or risk-tolerant investment portfolio. People who identified as Republicans were attracted to the conservative option, while Democrats were attracted to the more risk-tolerant options. However, when they were asked about the investment decision first, Republicans and

Democrats tended to choose the low-risk options with relatively equal frequencies.

Choices also tend to be supported by an illusion of understanding, Fox observed. People are overconfident in how well they understand how everyday objects, such as toilets or ballpoint pens, work. But when they are asked to explain in detail how such an object works, they realize that the mechanisms are more complicated and lower their self-assessed understanding of an object.

The same effect applies with public policies. For instance, when the Supreme Court upheld most of the provisions of the Affordable Care Act in 2012, more than three-quarters of Americans in a Pew poll expressed a perspective on whether they supported or opposed the ruling. However, barely half of them could correctly identify what that decision was. In another experiment, Fox and his colleagues presented individuals in an online sample with several policy issues—for instance, a cap-and-trade policy to curtail carbon dioxide emissions or a national flat tax. Respondents gave their positions and their level of understanding of the issue, after which they were asked either to give the reasons for their beliefs or to explain how the policy would have the desired effect. People who had to explain how a policy works subsequently rated their understanding of the issue as lower. They also described themselves as less likely to contribute to advocacy causes, especially those who were most extreme in their views.

Self-assessed understanding of scientific issues displays the same pattern. When people were asked to explain how carbon emissions affect climate change, they later rated themselves as having less understanding of the issue and more moderate positions—an effect not seen when they were asked just for the reasons for their beliefs. However, people whose illusion of understanding had been punctured were also less willing to support further research on the topic. "That's something we need to look at," said Fox.

Overweighting Marginal Views

Even on scientific topics surrounded by considerable consensus, such as climate change, low-probability events tend to be overweighted in making decisions, Fox observed. For example, if people are told that 10 percent of scientists believe *x*, many interpret this statement as meaning that *x* could be true or false, even though the position is held by only 10 percent of scientists. However, if people are led through the response of each scientist one by one—so this scientists believes *x*, this scientists believes *x*, this scientist does not believe *x*, this scientists believes *x*, and so on—they become more sensitive to the actual probabilities.

This finding has implications for how the media report on scientific results. A story that gives roughly equal emphasis to both sides of an issue may accentuate the public's bias to give undue credence to remote views.

People are undersensitive to probabilities in general, Fox said. For example, even as more scientists have become confident about the human contribution to climate change, public beliefs have not changed accordingly. In particular, media stories that devote time to marginal views have an exaggerated impact on the public's perceptions. The media need to think creatively about ways to communicate relative proportions in a way that people can absorb, he concluded.

What People Want to Know

As a further example of the ways many people are concerned about the motivations of scientists, Bill Hallman, professor and chair of the Department of Human Ecology at Rutgers University, described a set of open-ended interviews done with approximately 30 people on their perceptions of animal cloning. At the end of the interviews, the people were asked what else they would like to know about animal cloning research. Here are the top 10 questions in the order of how often they were asked:

1. Who is doing it?
2. Where are they doing it?
3. Why are they doing it?
4. What are the goals of this?
5. What is the status of the research?
6. What are the risks and benefits of the research, both to consumers and to the animals?
7. Who is monitoring and regulating the research?
8. How does it work?
9. What happened to Dolly?
10. Will we eat the food from cloned animals, and is it safe?

The most asked question involves who is doing the research. The third to the last question is how cloning works. Yet one of the first things science communicators want to do is explain the science, Hallman observed, even though this question ranks relatively low on the list of questions people ask. And if people do not get answers to the questions they ask most often, they are unlikely to absorb the answers to questions that are of lower concern.

Hallman also emphasized that people create mental models based on the information that is available to them. For example, science often works on objects and issues that are largely invisible to people, such as nanotechnology, climate change, or food safety risks. Science communicators need to understand the mental models people create of invisible things and speak in ways that connect with what people think they already know.

Improving Perceptions of Science

On the issue of warmth and competence, Hallman pointed to data suggesting that the perceived mismatch in a person who is coldly competent can exacerbate mistrust of that person. These data point to the need for scientists to develop greater social and emotional intelligence. Too often, said Hallman, scientists talk as if they are trying to impress other scientists, whereas if they put themselves in the position of a young student or a member of the general public, they could connect with what their audiences already know.

The media, which relentlessly stereotype scientists as cold, are part of the problem. But scientists also need to work against the stereotype. "We need, as part of our science education, to teach people how to tell a story, with a beginning, a middle, and an end, to tell a joke that's actually funny, and to take a joke when it's warranted." Standard resumés contain plenty of information on expertise and competence but very little on social abilities, despite the importance of these attributes.

The Problem of Persistent Minorities

One of the issues discussed during the question-and-answer period was the general reaction to individuals or small groups who insist that they are right and everyone else is wrong—what Fiske termed "persistent minorities." People often ascribe credibility to such individuals because they are standing up to everyone else and not caving in. "It's like the holdout in a jury," Fiske said. "That person has to be really motivated to resist all the other people on the jury." Because of this perception, people may favor a minority perspective even if it lacks credible evidence.

Hallman also brought up the costs to scientists when they claim to know what is going on and subsequently are shown to be wrong. A better strategy is to say what is known now and how likely that observation is while also describing what is being done to reduce uncertainty. In that case, if a scientist turns out to be wrong, people are more forgiving.

Fiske agreed that scientists do not have the right to tell people what to do. They have to provide information and talk about consequences in the most accessible way possible. Also, people need solutions, not just problems. If a scientist simply makes people afraid, they will avoid the topic, since they will assume that they cannot do anything about it.

Finally, Fox noted that stories can be more successful with a lay audience than data, since people's emotions are driven more by stories than by statistics. "A lot of science can be communicated in stories," he said.

COMMUNICATING UNCERTAINTY

To provide answers to the questions that decision makers ask, science communicators need to be able to communicate uncertainty, said Baruch Fischhoff, the Howard Heinz University Professor in the Departments of Social and Decision Sciences and Engineering and Public Policy at Carnegie Mellon University. Often that requires greater precision than scientists and communicators naturally provide. Behavioral decision science shows the problems caused by such ambiguity and ways to do better. It involves both analytical research, determining what people need to know, and descriptive research, making that information comprehensible. For example, studies of intelligence analysts have revealed the confusion potentially caused by verbal quantifiers such as that describing a Soviet attack on Yugoslavia as "a serious possibility." The U.S. intelligence community's current attempt to deal with this problem is seen in the standard explanation used in the *National Intelligence Estimate: Prospects for Iraq's Future: A Challenging Road Ahead* (FAS, 2007):

> When we use words such as "we judge" or "we assess"—terms we use synonymously—as well as "we estimate," "likely" or "indicate," we are trying to convey an analytical assessment or judgment. These assessments, which are based on incomplete or at times fragmentary information, are not a fact, proof, or knowledge. Some analytical judgments are based directly on collected information; others rest on previous judgments, which serve as building blocks. In either type of judgment, we do not have "evidence" that shows something to be a factor that definitively links two items or issues.

Although he appreciates the motivation underlying this clarification, Fischhoff said that "you would be hard pressed to figure out what they meant if you were actually a consumer of these documents."

Similarly, a recent analysis of 12 extreme climate events during 2012 stated that

> Approximately half the analyses [19 analyses by 18 different research groups on 12 extreme events during 2012] found some evidence that anthropogenically caused climate change was a contributing factor to the extreme event examined, though the effects of natural fluctuations of weather and climate on the evolution of many of the extreme events played key roles as well.

As Fischhoff observed, "You'd have to be a very select user audience to get the information you needed to make any decisions" from this analysis.

Questions Decision Makers Ask

The questions that decision makers ask of scientists can be divided into three broad categories, Fischhoff continued:

1. Whether to act,
2. What to choose, and
3. Whether and how to create options.

For each of these categories, communicators face both analytic challenges in extracting the information experts know and communication challenges in conveying that information with the precision that decision makers need.

Decisions about whether to act often involve determining whether a signal has passed a threshold. In this case, the analytical challenge is determining how much scientific information is available and translating this knowledge into an action threshold. Signal detection theory studies how well experts can discern a signal and how their reporting balances the costs and benefits of correct and incorrect responses. For example, Mohan et al. (2012) have looked at the decisions that physicians at regional or primary care hospitals make about transferring a patient to the emergency room at a tertiary (major) medical center hospital. Their analysis found great variability among physicians in determining when a patient meets the guidelines for transfer. The communication challenge in such a situation entails describing the decision rule that physicians use and how good physicians are at following it. Such communications reveal how much these experts know and how appropriately they apply that knowledge.

The next category of questions involves deciding between two or more alternatives. Here, the analytical challenge can be expressed by a decision tree, which, again, considers the uncertainties and expected consequences of each choice option. For example, the Food and Drug Administration uses a structured approach to summarize the expected benefits and risks in its drug regulatory decision making (FDA, 2013). FDA's approach considers the evidence, uncertainties, conclusions, and reasons associated with five decision factors. Those uncertainties may arise from variability in the observations, the internal validity of studies, their external generalizability, and the quality of the underlying science. Patients and physicians then can use this information in making their own decisions about a treatment.

The communication challenge in conveying such information includes explaining studies' methodological flaws and hidden values. All analyses embody values that favor some interests. When transparent, those assumptions can be controversial. However, they are often obscured in the measures that scientists and analysts use.

The third class of decisions involves whether and how to create options. The analytical challenges in making these decisions include identifying the relevant expertise and assessing the uncertainties created by omissions in the analysis. For example, an analysis by Casman et al. (2000) looked at options for dealing with drinking-water contamination, such as sending out notices to boil drinking water. The system for preventing health effects is complicated, involving detection, management policy, the public health system, how the messages go out, whether people know how to boil water well enough to eliminate contaminants, and other factors. If the analysis left out any of these factors, it could be missing an important part of the problem. Without such a comprehensive view, decision makers cannot understand the uncertainties underlying their choices. One particular challenge for this class of decisions is conveying the implications of excluded information.

A Reluctance to Express Uncertainties

Experts are often reluctant to express their uncertainties, Fischhoff observed. Sometimes they see such efforts as misplaced imprecision. Sometimes they think they will be misunderstood. Sometimes they fear being punished by their organizations for their candor. And sometimes they are uncomfortable with the elicitation method, not knowing quite how to express themselves in the required form.

Fischhoff had three proposals for dealing with the reluctance to grapple with uncertainties. One is to create standard procedures for making and communicating decisions. With such procedures, people get used to thinking in a particular way, organizing their evidence, expressing their thoughts, and getting feedback on how well they have done. For example, standardization has helped FDA not only with its external communications but also with its internal communications among its own personnel.

The second proposal is to create a resource center to provide experts with publication-quality support in eliciting and communicating uncertainty. Such a center could provide quality assurance, take advantage of economies of scope, anticipate common problems, form trusted personal relationships, and stimulate basic applied research into the challenges associated with analyzing and communicating uncertainty.

The third proposal is to create shared understanding of the analytical approaches needed to characterize uncertainty. All communication begins with an analysis of what people need to know and of the associated uncertainties. If scientists understood analytical procedures, they could work more easily with those who do communication science research.

An orderly treatment of uncertainty would produce more useful science by addressing decisions makers' needs, Fischhoff concluded. At the same time, it would produce better science by encouraging disciplined reflection on the uncertainties in scientific knowledge.

Uncertainties in Communicating Uncertainties

Communicating about uncertainties has its own uncertainties, said David Budescu, the Anne Anastasi Professor of Psychometrics and Quantitative Psychology at Fordham University, in his comments on Fischhoff's presentation. Some communications inform multiple types of decisions. Audiences are highly heterogeneous in their knowledge bases and mental models. And audiences may lack sensible models of the sources, causes, types, and limits of uncertainties. Many people, for example, think all uncertainties are alike, which can be highly misleading.

Given these uncertainties, the communication of uncertainties should be judged not by an absolute threshold but by the demands of particular circumstances, said Budescu. Frameworks of communication need to be flexible enough to accommodate different kinds of decision cycles and various levels of knowledge about uncertainty.

As Fischhoff said, experts are sometimes afraid to be perceived as too imprecise. But imprecision is sometimes necessary. With Tom Wallsten, Budescu developed the principle that uncertainty should be communicated in a mode that matches the nature of the event and the sources of uncertainty. For example, it makes little sense to communicate precisely vague uncertainties about ambiguous events, such as "the chance of abrupt change in the climate in the near future is 0.128." This would be an overly precise estimate for such an ill-defined and ambiguous outcome. By the same token, it is suboptimal to communicate imprecisely precise uncertainties about unambiguous events, such as the "chance of drawing the queen of hearts from a full deck of cards is quite low." The uncertainty needs to match the nature of the event.

When publicly traded companies issue forecasts of earnings per share, they often use imprecise forecasts, Budescu noted. An analysis of more than 33,000 quarterly forecasts by almost 5,000 companies issued between 1996 and 2006 found that forecasts citing a range were more accurate than point forecasts (51 percent versus 24 percent). Furthermore, the receivers of such information tended to prefer broader estimates because they judged the forecasts to be more informative, more accurate, and more credible.

People often have clear expectations of the appropriate level of precision or imprecision to communicate, Budescu concluded, and they value communications that more nearly match their expectations.

Meeting the Needs of Decision Makers

Adam Finkel, executive director of the Penn Program on Regulation and a senior fellow at the University of Pennsylvania Law School, made four points in agreeing with Fischhoff's analysis. First, one type of uncertainty is generated by interindividual variability. Unlike uncertainty, variation is irreducible, and unlike uncertainty, it forces decisions about risks and benefits to individuals.

Second, simple innumeracy can be the cause of mismatches between how experts offer information and how the public interprets it. As an example, Finkel cited a *New York Times* article that reported the failure rate of in vitro fertilization to be 77 percent while not mentioning that the failure rate of natural conception is also almost exactly 77 percent.

Third, instead of producing a weighted combination of mutually irreconcilable probabilities, a better option is a solution-focused process of decision that uses risks and benefits to discriminate among choices. For example, taking the midpoint of the probability distribution for hurricanes forming in the Gulf of Mexico is suboptimal because it does not consider the costs of error.

Finally, the most valuable information usually concerns the specific uncertainties that are plaguing decision makers and reducing the expected regret of the best possible decision.

The Responsibilities of Decision Makers

Given these four points, Finkel emphasized placing more of the onus on decision makers to demand better information. Decision makers need to act when they are not satisfied with the information they are getting because it allows them to make only rudimentary decisions.

He also emphasized that, in the work that he does on cost-benefit analysis for environmental and occupational health, the only reason to act is risk, and the only reason not to act is cost. However, the estimation of cost may be the broken link in cost-benefit analyses. Risk is what will happen if a policy is not created or implemented. Cost is what will happen if a policy is created and implemented. Yet economists have devoted relatively little attention to what policies cost. They also tend to minimize the uncertainties in estimations of cost, interpreting error bars as indicating an unfinished analysis, said Finkel.

Finally, Finkel cited two problems with estimates. First, they are often based on the mean, but the distribution of events around the mean is rarely distributed on a Gaussian curve. Rather, the mean depends on the characteristics of the tails of the distribution.

Second, the public and decision makers often are bound to what Finkel called "estimaticles"—estimates that, in the words of William Blake, act as "mind-forged manacles." When fully informed, a decision maker may choose a different estimator than one imposed by an economist or may conclude that no single estimator tells the whole story.

The Meanings of Uncertainty

During the discussion session, one colloquium participant pointed to the difficulties surrounding the word "uncertainty," which can lead decision makers to delay deciding until the uncertainty has been reduced. In such cases, it would be better to talk about confidence, best scientific opinion, or best estimate rather than uncertainty.

Fischhoff responded that these terms need to be tested, as do the other aspects of analyzing and communicating uncertainty. Uncertainty has a precision that works in the scientific community, but it may not work with a particular audience. Finkel added that deciding not to decide is still deciding, which essentially leaves a risk unchanged. Budescu agreed that uncertainty can be used as an excuse for inaction, as has often been the case with climate change. But with climate change, the uncertainties involve the rate of warming, not whether warming is occurring. Thus, the magnitudes and roles of uncertainty in a model need to be carefully specified and explained.

The presenters also examined several possible roles of scientists in communicating uncertainties. Finkel pointed out that the perception of being apolitical is crucial for scientists. For that reason, scientists should not overstep their bounds. They can provide information, including information about uncertainties, but decision makers should be the ones making the decisions.

However, in some cases, being apolitical may not be the best way of effectively communicating scientific information, he continued. It is more important to be transparent about one's values. People need to be willing and able to explain what they are doing when asked why they are doing something.

In any domain, people have acquired knowledge in school, from the media, and from friends, said Fischhoff. The challenge for communicators is to find out what people already know and how they can be informed in useful directions. If people are motivated and respected, they can understand a lot. "I think you're often surprised how far you can take people."

Policy arguments should be about thresholds for action, said Fischhoff, so as to identify value judgments, thereby maintaining the credibility of science by focusing it on the evidence. If pressed, scientists can show how evidence and values might be integrated in statements such as, "this is what I would do if I were in your situation." In many cases, that may not even be necessary, added Budescu, if the information and uncertainties surrounding a decision are made clear.

People have different levels of risk aversion, Fischhoff noted in response to a question. But emphasizing these differences can move the onus from the expert who should be explaining the uncertainties or the alternative options to individuals who are not necessarily well enough informed to make decisions. Budescu added that risk aversion is more domain specific than many people assume, so it cannot necessarily be applied globally to perceptions of risk.

Examples of Success

When asked for specific examples of successful approaches to communicating uncertainty, Finkel cited the job the Environmental Protection Agency has been doing on the health effects of air pollution. The agency has been careful about trying to separate the statistical uncertainties from individual variability. It also has prepared documents for decision makers that express the center of gravity and where the extremes are rather than subsuming different types of uncertainty into a single measure.

Fischhoff cited good television weather forecasters. They know their audience, have gotten feedback on their presentations, and know how the weather plays out in people's lives.

Budescu urged the use of multiple methods to convey uncertainty, including visual depictions and text. People should have a choice of methods or be able to experience multiple methods simultaneously.

SOCIAL NETWORKS

The analysis of social networks preceded the development of electronic social networking, said Noshir Contractor, Jane S. and William J. White Professor of Behavioral Sciences at Northwestern University. But several factors have come together that make it possible to understand and use networks in new ways. The social sciences have made substantial progress in understanding why people connect with others. New analytic methods for analyzing network dynamics and confirming hypotheses have become available. The use of "big data" makes it possible to analyze very large networks connecting large numbers of people through different kinds of technologies and platforms. And the computational infrastructure now exists to develop hypotheses and conduct analyses.

Changing Behaviors

The challenge in science communication is *not* to find better ways of communicating facts to people, Contractor said. Overwhelming evidence indicates that people hold onto their attitudes and behaviors despite, not because of, facts. Other approaches, therefore, are needed to change behaviors.

One is simply to ask people to do something. Another is to tell people that they must do something or they will incur penalties. A third approach is to create incentives to do something. However, neither penalties nor rewards are guaranteed to create a system or culture where people routinely engage in a desired behavior, Contractor said.

The literature on social influence suggests more effective approaches to change behaviors. For example, the psychologist Robert Cialdini has laid out six key principles of social influence:

1. *Reciprocity*: People tend to return a favor.
2. *Commitment and consistency*: Once people have made a decision, they tend to stick with it.
3. *Social proof*: People tend to conform and do what other people are doing.
4. *Authority*: People tend to obey authority figures, regardless of the situation.
5. *Liking*: People are easily persuaded by people that they like.
6. *Scarcity*: Perceived scarcity generates demand.

All of these strategies can be helpful, said Contractor, but they are general strategies, and scaling them up can be a major challenge.

The Who of Social Influence

Scaling up science communication to reach large numbers of people requires leveraging three types of knowledge, Contractor said:

- Science about how social influence strategies can be effective,
- Science about who the touch points are in networks, and
- Science about strategic choices involving social media.

Focusing on the "who" of social influence rather than the "how" provides particular promise for scaling up science communication. For example, researchers have shown that ideas can be introduced by the mass media but spread to larger publics via opinion leaders. In addition, people are socially influenced by the people they know and trust when forming an opinion or engaging in a behavior. These observations become even more crucial as new social media platforms add to the firehose of information people receive.

The field of social network analysis has sought to identify trusted opinion leaders who can help disseminate information. For example, one aphorism describing social networks, Contractor said, is "it's not what you know, it's who you know." This description can seem disparaging, but success often depends on both what someone knows and who that person knows.

A second aphorism, which describes cognitive social networks, is "it's not who you know, it's who they think you know." People act on the basis of their perceptions, so the perception of being part of a social network can spur action.

A third aphorism, which describes knowledge networks, is "it's not who you know, it's what they think you know." People often act on the basis of stereotypes rather than factual knowledge about other people.

The final aphorism, which describes cognitive knowledge networks, is "it's not who you know, it's what who you know knows." In this case, social networks and knowledge networks are merged.

People use all four of these ideas of networks every day, Contractor said, though some people leverage particular networks more than other people do. Furthermore, all of these networks are becoming more complex. The result is a multidimensional network where some of the nodes are people, some are concepts, some are keywords, and some are hashtags. Communicators need to see themselves as embedded within these different contexts if they are to understand how to leverage networks.

A Strategy for Leveraging Networks for Science Communication

Contractor proposed a strategy for leveraging networks for science communication that involves three elements:

- *Discovery*: "If only we knew what we know."
- *Diagnosis*: Identify the touch points that can serve as a multiplier for scale-up of scientific communication.
- *Design*: Recommend strategies for selection of social influence, touch points, and social media to scale up scientific communication.

With regard to discovery, people often do not know what they know. As a result, they spend time reworking issues because it seems less expensive to reacquire knowledge than to access already acquired knowledge.

Diagnosis involves finding the key people in a network and using the best strategies to influence those people. Naturally occurring networks are not always efficient or fully functional. For example, disparities in information sharing via networks can lead to pockets of information haves and have-nots, thereby increasing knowledge gaps. Also, not all nodes are created equal. Some nodes are particularly well positioned to be touch points or to serve as a network multiplier in a scale-up effort.

Design involves selection of the right social influence strategy, the right touch points, and the right social media channels to optimize the speed and coverage of communicating scientific information to publics or targeted audiences. In particular, the science of networks can help reveal which touch points are most likely to serve as multipliers for a scale-up effort.

Being Influenced by Others

In a famous experiment described in 1955, the psychologist Solomon Asch asked undergraduates to say which of three lines drawn on an index card was the same length as a single line drawn on a separate card. The task was so simple that no one got the answer wrong when they were asked the question at the beginning of the experiment. But if the students were first exposed to a series of other participants consistently choosing the wrong line, 37 percent chose that line as well.

This experiment is a classic demonstration of how people often conform to behaviors that they know are wrong, said Katherine Milkman, the James G. Campbell Assistant Professor of Operations and Information Management at the Wharton School of the University of Pennsylvania. Other research has shown that people conform even to behaviors that are only described to them. For example, if college students are told that 75 percent of their peers engage in a certain behavior, such as safe sex or moderate drinking, they are more likely to engage in that behavior themselves. As another example, messages about a household's electricity usage that include information about the usage of neighboring households can reduce overall electricity usage over long periods of time. Even attributes as fundamental as obesity, smoking, happiness, and loneliness can be spread through social networks, Milkman said.

New social networking technologies raise intriguing questions about observance of social norms. For example, research has shown that Facebook users were more motivated to vote when they got messages about friends of theirs who had voted. Leveraging such interactions poses both opportunities and risks for science communication, Milkman said.

Risks Posed by Social Media

As an example of the risks posed by social media, Milkman noted that social networking technologies can support herding, where people associate only with like-minded compatriots. On the flip side, herding may be used for constructive purposes. Technologies make it possible to track, for example, a snowball effect taking place on the web, so if a video has gone viral and is communicating misinformation, people can take countervailing steps to counter it.

As another example of possible risks, Milkman pointed out that efforts to help someone within a network can hurt someone who does not have as strong a connection to the network. For example, women and minorities who do not have the same strong network connections as men or majority populations are less likely to receive the same recommendations and favors.

The Future of Social Media

In the past, new communication media such as radio, the telephone, and television, each of which had sweeping effects on society, took a couple of decades to develop and deploy at scale, said Deb Roy, a tenured professor at the Massachusetts Institute of Technology and chief media scientist at Twitter. Today, the convergence of communication media and computers has made it possible to develop new communication mechanisms within weeks. Once an idea has been developed, software can quickly be written for hardware that already exists. The development of communication platforms has gone from what Roy called a solid phase to a liquid phase.

In such circumstances, envisioning the state of communication media 5 to 10 years from now is much more difficult than it was in the past. Cultural adaptation, as much as technological capability, is what determines which communication platforms spread and how fast they do so.

Changes in the Culture

In response to a question from a participant about the factors that keep scientists from using new media, Roy responded that the 140-character limit to tweets is not a serious limit, because people can always tweet again. But it is a format that works well for some kinds of communications and not for others, he said. For example, tweets can point to new scientific papers, intriguing observations, and other points of interest. As another example, the Sackler colloquium itself generated a healthy stream of tweets, Roy noted. When one of these is retweeted, it can end up in the timeline of someone unaware that the colloquium was occurring. The result is an ad hoc dynamic social network that can amplify messages and create long-lasting links.

New media may initially be seen as lightweight or trivial, but they evolve over time, Roy said. Today, scientists are using new media successfully, and new uses will continually be discovered. Contractor added that the culture has changed in institutions in the past few years, and people who know how to use these tools and use them effectively are now being rewarded. For example, more portfolios submitted to tenure and promotion committees include metrics involving new media. Among younger students and graduate students, many are adept users of Twitter and other new media.

Media Adaptation

Contractor also noted that when a new communication medium is developed, people tend to think that it will democratize access and bring back the public square. But history often indicates otherwise. For example, new media can help people connect more strongly with others like themselves, creating echo chambers for opinions. Even in science communication, new media can create an interest in talking with like-minded others. In such cases, research can point to people who can serve as bridges or brokers between groups.

Network dynamics also can vary from place to place, Contractor said. People in India use Twitter and Facebook frequently, but not for professional work. Most companies in China frown on their employees using LinkedIn because it indicates that an employee is looking for a job. Distinct norms emerge in different places and evolve over time.

Contractor added that when a new medium is developed, the existing media need to adapt if they are not to be displaced. Radio replaced the newspaper as the main place where people got their news, and radio in turn was replaced by television as the major means of news dissemination. As the communication ecosystem continues to explode, all media will need to adapt to continuing change. Furthermore, many communication media remain in what some have called permanent beta, where they continue to change and evolve.

Don't Believe Everything You Read on the Web

During a follow-up conversation on the risks posed by new communication technologies, Contractor pointed out that echo chambers have a natural tendency to form in new networks. Research may suggest ways to open up the dialogue, even if doing so would push people out of their comfort zone.

Contractor also observed that new media make it possible to target specific audiences with specific messages to accomplish a common good, even though people are reached in different ways. Roy agreed that it is easier to deliver a specific message to targeted groups, but this can make it difficult to reach large numbers of people because it is hard to differentiate messages for so many different groups.

As another potential risk of new media, Contractor pointed to the "dark web," where people can mobilize using new media without being publicly visible. Even Facebook has secret pages that invite large numbers of people by invitation only.

Roy acknowledged the risk of misinformation spreading through new technologies. Information on the web can be difficult to trace back to its source, reducing certainty about the veracity of that information. Better education and better tools are needed to help people make sense of what they encounter through social media. As Contractor reminded the participants, a useful reminder is the quotation "Don't believe everything you read on the web.—Abraham Lincoln."

Measures of Success

Regarding metrics of success, the particular people being reached may be more important than the absolute number, said Roy. In addition, the content of the messages that flow through a network can be used to measure the effectiveness of dissemination.

Tools are being built to assess who is reached, Contractor observed, but they are in their early stages. For example, two people with the same followers do not necessarily increase the distribution of a message as much as two people with different followers. Similarly, if a person's followers have many followers themselves, that person may be a more effective disseminator of a message.

SCIENCE COMMUNICATION AS POLITICAL COMMUNICATION

Science is encountering politics more and more often, and the trend will not change anytime soon, said Dietram Scheufele, John E. Rose Professor in the College of Agricultural and Life Sciences at the University of Wisconsin-Madison. The societal applications of modern science are inherently political issues, whether the issue is stem cells, climate change, obesity, or synthetic biology. This is the case despite the fact that many members of the public know relatively little about these issues. For example, when a sample of the public was asked whether it was true or false that the Obama administration had recently banned all research on synthetic biology, about a third were able to provide the correct answer, 12 percent thought it was true, and 55 percent did not know.

An idealized model of society holds that events occurring in the arenas of politics, science, economics, or other societal domains have a direct influence on perceptions of reality and public opinion. But very few people can observe these events directly, Scheufele noted. Most perceptions of reality are mediated, usually through the media. The information transmitted through the media is in turn selected through a process known as agenda building in which people negotiate the content of media messages. For example, corporations work to push certain content, while scientists and their institutions write press releases to get their stories covered. In this negotiation, science is only one of many voices in society.

Research has shown that media coverage of particular science issues increases when politics become involved in the issue, and media coverage spikes when an issue becomes controversial. Thus, science tends to get covered by the media when politics become involved. Furthermore, people tend to remember and use information that they get from the media—a phenomenon known as media priming. If an issue is ignored by the media or by the people the media is covering, it will not become salient in public perceptions.

Framing

Information derived from science is often ambiguous, Scheufele observed. Carbon nanotubes may cause cancer because they behave similarly to asbestos fibers, but carbon nanotubes are also important components of many types of materials and equipment that may allow for the early detection of cancer. For a lay audience with no training in nanotechnology, this is an ambiguous stimulus that could be interpreted one way or the other.

All perception depends on the context, especially for ambiguous stimuli. The framing of information therefore shapes how people think about that information. Sometimes framing has a partisan motivation, but in most cases it is simply a tool for information processing to help people determine why an issue is important and how to think about it. Framing reduces ambiguity by contextualizing information, and it is most successful if it resonates with an underlying schema.

There is no such thing as an unframed message, Scheufele said. Even in an entirely professional setting, such as a scientist's grant proposal to the National Science Foundation (NSF), the use of framing is inevitable.

Making Sense of Science

Researchers have developed several models of how public opinion is formed, both at the individual level and at the societal level. For example, the expression of opinions can vary depending on individual characteristics. Males are more likely to speak out in controversial situations than females, and young people are more likely to speak out than older ones. But expressing opinions also depends on the social environment. If a particular opinion is not favored in a society, then people are less likely to express that opinion, which in turn leads other people not to express that opinion. The result is a spiraling process in which particular opinions can become dominant because people in the minority are much less likely to express their views.

Social norms campaigns use the same logic to shape opinions. For example, when hotel guests see a sign saying "75 percent of the guests who stayed in this room reused towels," they are more likely to reuse their towels than if they see a sign saying simply "75 percent of guests reused towels" or "You can show your respect for nature and help save the environment by reusing your towels."

Motivated Reasoning

People process information based on their beliefs, identities, and ideologies. Studies of this process of motivated reasoning are not new but have seen a recent renaissance, said Scheufele. Motivated reasoning functions both through selective exposure to information and through the interpretation of that information. When people firmly believe something, they are more likely to seek out new information that conforms with that belief. They also are more likely to question information that does not fit with that belief. However, today's targeted media environment is making it more difficult to be exposed to debates and the other side of the issue. Even newspapers soon could be customized to give people only information they want.

Ambivalence about a topic makes if more likely that people will engage in both sides of an argument. Also, if people are in groups that disagree with them so that they have to justify their opinions, they are more likely to process information carefully.

As a final example of this research, Scheufele pointed to work in political communication on why having a heterogeneous network that exposes a person to different opinions is correlated with having more factual political knowledge and greater public participation. "Disagreement is good for us for a whole variety of reasons and actually forces us to think through some of these things more carefully."

Next Steps

Separating the social context from the science is critical for the scientific endeavor but is dysfunctional for science communication, Scheufele noted in closing. Whether scientists like it or not, new science will be debated in a complex media environment where individual predispositions and societal expectations loom large.

In such an environment, systematic efforts are needed to increase citizens' ability to find scientific information in increasingly fragmented media environments, connect science to their daily lives, and process information accurately. It is not just about getting information to citizens but about helping them get it right, Scheufele said. There are no easy answers for how to do all of these things, but the questions can be addressed through empirical research.

Problems with Science Communication Research

Science communication research is not necessarily a cumulative scientific enterprise that gives practical guidance on difficult issues, said Patrick Sturgis, professor of research methodology from the University of Southampton. It is better characterized as a loose assemblage of interdisciplinary frameworks and approaches. As a result, it will not necessarily provide easy off-the-shelf solutions to the long-term problems of science communication.

Sturgis described several of the key problems associated with science communication. One problem is that small variations in the wording of questions make big differences in the answers obtained. The wording of questions can shift apparent public opinion from a minority to a majority position, which is "worrying if what we think we're doing is measuring something real and concrete."

Another problem is that people are willing to provide opinions on nonexistent issues. Polls demonstrate that many Americans are in favor of the monetary control bill, the agricultural trade act, and so on, but these pieces of legislation do not actually exist. In Britain, people were even more strongly in favor of nonexistent bits of legislation, Sturgis noted. If people are willing to offer opinions about nonexistent issues, it must lead to questions about the robustness of opinions measured on genuine questions of public policy.

There, problems are particularly acute in the area of science and technology because many people have heard little or nothing about key areas of research and application. A 2012 Wellcome Trust survey of adults in Britain, for example, found that over half of the population had never heard of the term human genome.

Science communication research has a valuable space in describing the shape of public understanding and preferences, but it is on shakier ground when it comes to explanatory accounts. Because most data are observational and nonexperimental, multiple accounts of how they arose are often equally plausible. For example, accounts of motivated reasoning can be equally well explained through an "enlightened preference" framework. A recent study found that greater knowledge about genetics does not necessarily lead to greater approval of the use of genetically modified crops. It does for people on the left of the political spectrum, but for people on the right side, greater knowledge leads to less approval for the use of such crops. This can be interpreted as a case of enlightened preference rather than motivated reasoning, in which an increase of knowledge enables people to connect their core values to their policy preferences. More generally, it points to the difficulty in making causal inferences from observational data.

Despite this somewhat pessimistic perspective for the ability of science communication to deliver, Sturgis concluded by pointing to the "huge amount of benefit" that can be derived from the interdisciplinary field of science communication. However, social scientists need to avoid overpromising what the field can deliver, a trap that bench scientists are often accused of falling into themselves. They also need to avoid the implicit promise that they can provide insights that will enable scientists to get the public on their side for the latest favored technology.

Microtargeting and Counterframing

Kathleen Hall Jamieson, Elizabeth Ware Packard Professor of Communication at the University of Pennsylvania's Annenberg School for Communication, made four points in commenting on Scheufele's presentation. First, microtargeting—repeating a tailored message through multiple channels, including social media channels, to a small target audience—is becoming increasingly common. For example, during the 2012 presidential campaign, Republicans targeted groups in coal states to receive targeted messages about "the Obama war on coal." This approach can have a powerful influence on targeted groups by creating coherent arguments and supporting strands of evidence. At the same time, microtargeting works against the kinds of exchanges across ideology or perspectives that can expose people to new information. For example, microtargeting strategies rarely raise the microtargeted message to a higher level because doing so would make it susceptible of being identified as inconsistent.

The second point Jamieson made is that presidential debates, in contrast to microtargeting, function well in American society by helping people learn about the candidates and issues. They feature both sides of the campaign and are watched by large numbers of people. As a result, they promote postdebate discussions that at least have the potential to intersect with non-like-minded networks. Even if they may not be persuaded, people are at least going to hear the other side. The success of presidential debates prompts the question of how to create more mass exposure experiences featuring exchanges about the things that matter to public policy.

Jamieson's third point is that framing does not have much of an influence when counterframing also exists. Most experiments that show framing to be powerful do not include counterframing. Where counterframing exists, as in highly partisan media environments with high levels of microtargeting, framing's influence is reduced.

Jamieson's fourth point involved the credence given to minorities that resist majority opinions. Historical examples such as Galileo indicate that minority resistors can be correct. As a result, when people argue aggressively and persistently for a position, they tend to be taken seriously.

Changes Due to Social Media

The potential of social media to change science communication was a prominent topic of discussion during the question-and-answer period. Scheufele said that social media and other online media can give scientists a voice they have not had before by enabling them to present issues in the ways that they desire. But messages disseminated through social media can have a widespread effect only if they can take advantage of a multiplier effect.

Jamieson added that the linking and alerting functions of social media potentially can be used to motivate individuals to gain deeper knowledge about an issue, including greater ability to argue and counterargue an issue. This is a very rich area for scholarly pursuit, she said.

Sturgis, however, was pessimistic about social media's ability to better connect scientists to parts of the public that are not usually reached. People select into social media channels, and if they are not interested in information they tend to be put off by it. Social media have extraordinary potential for communication among scientists, "but I very strongly doubt that it will open up a new channel of unmediated communication between scientists and lay public," he said. Sturgis also noted that the entire public, not just the users of social media, need a voice on matters that are consequential for public policy.

Social media opinions are not proxies for public opinion, Scheufele agreed. But social media can point to emerging themes, which in turn can suggest ways to start conversations and promote engagement.

From Mass to Micro Media

In many ways, said Scheufele, the idea of a mass medium is declining. The media environment is much more diverse than in the past, and people have much more access to this diverse environment. The problem is that people do not have a motivation to seek out information that does not accord with their preconceptions. Even on social networking sites, people are far more likely to have friends with similar views than different views.

But Jamieson countered that the mass media era is not yet over. Even if a transition is under way to other sources of information, much of the information on the Internet originates in a mass media channel of some sort.

With any medium, people need to be able to engage in a search for accuracy, Scheufele observed. Part of the motivation for this search comes from being socially accountable. Part comes from a willingness to acquire information from heterogeneous networks. Rather than providing more information to counter misinformation, can people be taught how to make a more effective cognitive investment in assessing the accuracy of information, so that they can make more enlightened choices even if they do not fully understand an issue?

The Limits of Public Opinion Polls

During the question-and-answer session, Jamieson also discussed some of the problems with polls of public opinion. They often force people to make choices when in fact their opinions are unformed. The survey research system needs to spend more time figuring out what people actually know, which would indicate what they need to learn to enter into a discussion.

Reporters used to report polls very uncritically, Jamieson pointed out. Now their stories often include such information as margins of error, response rates, and the existence of framing effects. This is an example of how the education of journalists can help protect the integrity of journalism.

LESSONS LEARNED

Fischhoff concluded the first day of the colloquium by identifying four points that he considered the highlights of the day.

First, the science of science communication has ways of better understanding other people. Social scientists have produced tools to learn about individual interactions. These tools make it possible to avoid thinking that public opinion is refractory to any sort of external criticism. As Susan Fiske said earlier in the day, Fischhoff noted, people are not idiots. The science of science communication can help figure out what they are thinking and doing. In the process, scientists and science communicators can reflect on themselves and broaden their own perspectives.

Second, science communications can be improved. Even though there are no magic bullets, a base of knowledge exists that scientists and science communicators can use to avoid things known not to work and to use approaches that have a better chance of working. Outcomes then can be monitored to increase the rate of success.

Third, changes in social and intellectual organizations are needed to improve science communications. For example, integrated multidisciplinary teams are needed so that people can work together on complex multidisciplinary problems, which will require that institutions support scientists to do this kind of work.

Finally, scientists and science communicators themselves need to be willing to create the new forms of interaction needed to solve the problems they face. They need to engage in the kinds of research that will bring disciplines together and avoid fragmented and potentially misleading efforts.

2

Science in a Time of Controversy

Scientists often struggle to communicate their work, said Ralph Cicerone, president of the National Academy of Sciences, in his introductory remarks on the second day of the Arthur M. Sackler colloquium on The Science of Science Communication II. Even clear descriptions of evidence based on careful experiments, observations, or calculations do not always get through to many audiences. As a result, scientists have realized that they need to learn new ways of communicating scientific information to nonscientists.

In recent years the National Academy of Sciences, National Academy of Engineering, Institute of Medicine, and National Research Council have grown increasingly concerned about the communication of science-related issues to the public. As a consequence, the institutions have been inviting social scientists to meetings to learn more about the challenges of science communication. The institutions have learned about how people form beliefs and attitudes and why scientists sometimes get caught in the middle of political, economic, or moral disputes. They have learned about the economic and social factors that can shape science communications and about the potential of social networks. Those interactions led to a more systematic engagement when the Academy hosted the first Sackler colloquium on the Science of Science Communication in May 2012.

Like the first colloquium, the second was designed to push people in "uncomfortable ways," said Cicerone. Social scientists were asked to bridge the gap from their controlled studies to the complex world in which communication actually takes place. Science communicators were asked to augment their own professional judgment with scientific evidence about how they communicate. Subject-matter experts were asked to listen both to social scientists and to practitioners while making certain that they get the facts right. The object of the colloquium was to foster innovative thinking and fruitful new collaborations through interactions that would not have occurred otherwise.

RESPONDING TO THE ATTACK ON THE BEST AVAILABLE EVIDENCE

Kathleen Hall Jamieson, Elizabeth Ware Packard Professor of Communication at the University of Pennsylvania's Annenberg School for Communication, began her keynote address on the second day of the colloquium by looking at two broad communities that are involved in science communication.

The scientific community is part of an expert or elite community responsible for knowledge generation. This community also includes institutional entities such as the Bureau of Labor Statistics and the Congressional Budget Office that use the best available methods to generate knowledge, internally critique their methods to improve them, and police what it is that they communicate. This community seeks to make sure that it does the best it can to communicate what is knowable with the best available evidence. It usually does not achieve complete certainty, but it also seeks to communicate the levels of uncertainty associated with knowledge.

Journalists are part of a community that does not generate knowledge; rather, it uncovers and transfers knowledge that already exists. Journalism, too, is responsible for being transparent, for disclosing how it does what it does, and for policing itself. When it makes mistakes, the journalistic community, like the scholarly community, is expected to correct the record.

Both communities, when they perform their functions well, inform the policy-making community. In the process, they are able to hold the policy-making community accountable for its actions in relationship to the knowable. This model is not a completely accurate description of reality, Jamieson acknowledged, but it provides a framework for analysis and discussion.

Correcting Mistakes

The expert community's policing process requires that the public understand how it knows what it knows. When the expert community certifies that it knows something, the public should be confident that it does.

Sometimes the scientific community gets something wrong. For example, the peer-reviewed 1998 article by Andrew Wakefield that associated vaccination in children with a pervasive developmental disorder was a highly consequential case in which peer review failed and the expert community communicated something that it should not have communicated. The journalistic community also was at fault in this case by not uncovering the overall scientific consensus. Most journalists ran with the story, but one journalist began to investigate improprieties in the research that eventually led to its being discredited. When the scientific community learned of flaws in the evidentiary chain, it retracted the article. In this case, both communities eventually acted as they should have. People were still hurt, but fewer people were hurt than would have been if the two communities had not eventually maintained their standards of self-correction.

One problem with self-correction is that it can fan the suspicions of those who believe that these communities are inept, duplicitous, or partisan, Jamieson observed. To prevent such suspicions from undermining the integrity of these communities, they need to frame the correction in such a way that the public understands that these communities are acting as they should. If corrections were common, the functioning of the communities could be questioned, but most of the time they get most of what they do right. Some percentage of the population will never be persuaded on particular points, Jamieson acknowledged. But democratic systems have mechanisms to decide when sufficient numbers of voters agree to take action.

Factors that Undermine Credibility

However, this model of knowledge generation and dissemination can be undermined, Jamieson continued.

First, the custodians of knowledge can be challenged. For example, nonpartisan institutions such as the Congressional Budget Office can be attacked as not using the best available evidence and methods but instead articulating a partisan position. Similarly, politicians may attack journalists as partisan when reporting infringes on their ability to construct a reality for the electorate.

Second, individuals have a tendency to see evidence through a partisan filter. When a partisan perspective is applied, evidence may be used in a partial fashion, not in a way that represents the best use of all the available evidence.

Effectively responding to such attacks on the use of the best available evidence requires acknowledging the two personae functioning in every communication, said Jamieson. The first is the communicator; the second is the intended audience. Both have obligations that they bring to any exchange.

The scientific community needs to be credible, impartial, and respectful. Polemics are outside the engagement process that is appropriate either for journalists or for the expert community. The scientific community therefore needs to view its audience as intelligent, thoughtful, and worthy of engagement. Scientists do not have the responsibility to make policy decisions, but science has the potential to establish a context within which people can decide to act. It can lay out the consequences of doing something and the mechanisms that produce those consequences. In this way, it can enable people to make decisions based on the best knowledge backed by the best available evidence.

The scientific community also needs to find common ground with the audience it is trying to reach. Effective communication is built on shared premises or assumptions. An audience has to invest meaning in a communicative exchange. This meaning exists at the intersection of a text, a context, and a receiving audience.

Finally, scientists need to try to share knowledge, not impose it. They can inoculate against opposing claims by explaining what is known and what is not known.

The Attributes of Science Communications

The integrity of all evidence needs to be rigorously scrutinized. But science communication also needs a way to convey the existence of a consensus, Jamieson observed. Communications can emphasize that the experts who have formed the consensus have been right in the past. They need to employ a voice that is credible, impartial, and trustworthy. They need to communicate that scientists care about this issue and problem but also that they care about the integrity of their methods and in providing information that is as accurate as possible given the existing state of knowledge. Sources of funding need to be disclosed so that people can judge the effect that a funding source might have on the credibility of scientific results.

Journalists also have a responsibility of determining when a consensus does not exist. When journalists do not scrutinize the available evidence, they run the risk of conveying misinformation. Journalists can err both by failing to report on consensus and by assuming a consensus that does not in fact exist.

Besides conveying the presence or absence of consensus, science communications need to be nonpartisan, so that they can counter partisan filters. They need to provide a construction of reality that lets people understand what is known and how it is known. And science communications need to use a voice that conveys respect for the audience.

Rewriting Headlines

Jamieson used several examples to illustrate her points, one of which involved the following 2013 headline from the *New York Times*:

> "Arctic Ice Makes Comeback from Record Low, but Long-Term Decline May Continue"

This headline frames the issues discussed in the article in a highly misleading way, Jamieson contended. It fails to point out that the long-term trend in Arctic sea ice has been steadily downward. It cites further declines as possible but uncertain. As Jamieson said, an asteroid "may" destroy the National Academy of Sciences building, but it is highly unlikely.

On the Fox News website, this news was framed as follows:

> "Arctic sea ice up 60 percent in 2013"

The article went on to note that the increase is "a dramatic deviation from predictions of 'an ice-free Arctic in 2013.'" The article also said, "The surge in Arctic ice is a dramatic change from last year's record-setting lows, which fueled dire predictions of an imminent ice-free summer."

However, the article also cited data from the National Snow and Ice Data Center of Boulder, Colorado, showing the overall trendline in sea ice as going down. This point of common ground can be used to forward, deeply and respectfully, an alternative argument—that the amount of sea ice has been declining over time. Furthermore, this argument can be supported by visual demonstrations and additional data to reinforce the concept that the trendline is down. In addition, vivid images, clarifying metaphors, and evocative narratives can be used to reinforce the overall message—such as the idea that the reduction of sea ice is dramatic enough to see from the moon, or reports of changes in polar bear behavior and mortality.

Conveying a consensus to the public requires time, education, and breaking through partisan filters, Jamieson concluded. When scientists are positioned as nonpartisan, evidence is more likely to be heard and scientists are less likely to be seen as polemicists or persuaders.

PUBLIC ATTITUDES, STAKEHOLDER PERSPECTIVES, AND THE CHALLENGE OF "UPSTREAM" ENGAGEMENT

For some important issues in science and technology, public engagement and communication need to move upstream to much earlier in the research and development process, said Nick Pidgeon, professor of environmental psychology at Cardiff University in Wales. The objective is not just to persuade someone that a new technology should be accepted. Rather, a more discursive and two-way approach to public engagement can foster better overall decision making.

In the case of nanotechnology, for example, the Royal Society report *Nanoscience and Nanotechnologies: Opportunities and Uncertainties* (Royal Academy of Engineering, 2004) made the point that early engagement and dialogue can achieve several critical ends:

- Incorporating public values in decisions,
- Improving decision quality,
- Resolving conflict,
- Establishing trust and legitimacy, and
- Education and information.

Upstream engagement also has several obvious difficulties, Pidgeon continued. People are likely to know less about a technology at an early stage of research. Mental models of risk processes are likely to be absent or ill formed, with analogies instead serving as a proxy of risk. Both the future course of the science and potential regulatory needs or gaps will probably be uncertain, and the promoters and detractors of a technology are likely to issue both hype and dystopian narratives.

Geoengineering

All of these difficulties are apparent when considering geoengineering—the intentional manipulation of the Earth's climate to counteract warming or other aspects of climate change. Modification of the climate has been discussed for decades, but geoengineering as a way to counter climate change has been seriously discussed for only a few years, and scientists are deeply conflicted about it. It involves reflecting solar radiation back into space to lower global temperatures or removing carbon dioxide from the atmosphere, either of which would require engineering projects of immense scope. Neither the feasibility nor the full consequences of such methods for geoengineering are yet known.

In 2010 the equivalent in the United Kingdom of the National Science Foundation (NSF) commissioned the Stratospheric Particle Injection for Climate Engineering (SPICE) project. Conducted by a large consortium of engineers and scientists, the project explored the possibility of delivering reflecting aerosols through a 20-kilometer pipe tethered to a giant weather balloon. The project involved laboratory experimentation, modeling, and background review of the project and its possible impacts and risks. The project also included a proposed field trial—a small-scale 1-kilometer mock-up with a small balloon spraying water to answer some basic engineering questions.

The project was approved by two university ethics boards on the grounds that it did not jeopardize human health, interfere with animals, or have any detrimental effect on the environment. However, the reaction of the press and of some nongovernmental organizations was intense. Given the sensitivities of the technology and its implications, the research governance protocols that allowed it to be approved have to be questioned, Pidgeon said.

A Framework for Responsible Innovation

In response to the controversy, the SPICE researchers were asked to address five criteria before the pipe and balloon test could go ahead. One was to identify mechanisms to understand wider public and stakeholder views regarding envisaged applications and impacts of the experiment. Pidgeon's team then was asked to design a protocol that would enable members of the public who knew very little about the technology to form a considered opinion on whether the field trial should go ahead.

Developing such a protocol was immensely challenging both in conception and in methodological terms, said Pidgeon. It required intensive piloting, extensive engagement with the SPICE team and other geoengineering experts, and input from a stakeholder advisory panel. Three two-day workshops were conducted in different British cities, with 10 members of the general public selected to participate in each workshop. The aim in each workshop was to bring participants up to speed on the science and ethics of geoengineering and then to solicit their views. Methodological considerations in holding the workshops included framing the issue and the materials and experts to be employed. For example, workshop participants needed to be exposed to different framings of the issues involved (on both technical and ethical questions) to avoid presupposing their positions.

A particular methodological consideration was which people should be included in the workshops. There is a difference between an audience chosen essentially at random, such as the jury approach eventually adopted in this case, and an audience with a preexisting interest in a question. People with a preexisting interest can have a different set of attitudes prior to an engagement, yet they can be just as important in deciding an outcome as people without a preexisting set of attitudes. Another question for future dialogues around geoengineering involves whether participants should be from developed countries or from the less developed countries that are likely to suffer the most severe immediate consequences of climate change or unintended adverse impacts of geoengineering. These and other questions regarding the make-up of participants in such exercises are still being debated.

Following the workshops, very few participants wanted to rule out the 1-kilometer test, Pidgeon reported. They felt that it was a good thing for scientists to explore the topic, even though their views on the use of stratospheric aerosols were very negative, since people are disturbed by the thought of interfering with natural systems on a planetary scale.

In controversial areas, Pidgeon concluded, scientists and science communicators need to respect the views of the public if science is to progress. In addition, decision making over issues that will affect our lives in the future requires an emotional commitment as well as the analytical weighing of costs and benefits.

Upstream Complications

Upstream dialogue is extraordinarily important, agreed William Hallman, professor and chair in the Department of Human Ecology at Rutgers University, in his comments on Pidgeon's presentation. But scientists and science communicators need to be very careful with what they do upstream, "because what we put in the water upstream we end up drinking downstream."

One issue is that consensus needs to exist that a particular problem is worth discussing, Hallman said. Once this consensus exists, discussion can proceed on whether a particular technology is the right way to solve a problem.

Upstream dialogue is also complicated by the fact that most members of the public will know very little about the topic (though this often will not prevent them from stating an opinion). Those initiating the dialogue therefore need to be very careful about what they bring to the discussion.

Downstream Consequences

Rick Borchelt, director of communications and public affairs for the Department of Energy's Office of Science, also agreed that upstream engagement is laudable as a democratic ideal. In practice, however, it is fraught with potential problems.

Climate change is distinct in posing a dire problem that needs to be solved. But other technologies do not necessarily present problems in their earliest stages of development. The question therefore becomes whether upstream engagement is generally applicable to many of the issues that interest science communicators.

In addition, upstream engagement runs the risk of inciting others to develop counternarratives that might not have existed if the engagement were not performed. If the objective of upstream engagement is to fend off future controversy, the question becomes whether the efforts should be done without reference to the possible controversy or as a fully transparent exercise. One possibility is that engagement will create an arms race downstream over issues. In general, the upstream environment is rarely free of the controversies that upstream engagement presupposes, Borchelt said.

In general, the nature of the engagement process is critical. For example, scientists need to both provide information and gather information through listening. If they are not prepared to listen as well as talk, they should not be going into an engagement opportunity.

THE BENEFITS OF EXTREME SIMPLICITY IN COMMUNICATING NUTRITION SCIENCE

For a guideline to change behavior, it has to be memorable and actionable, said Rebecca Ratner, professor of marketing at the Robert H. Smith School of Business at the University of Maryland. An exception involves guidelines so complicated that they cannot be easily remembered, in which case a checklist can be an effective way to influence behavior. But in general, a guideline must be remembered by the target audience, and they must be able to do what it recommends.

Nutrition is an area where guidelines are useful since people generally cannot consult a guide every time they make a decision about what to eat. For example, the food pyramid, which was developed in 1992, called for people to eat 6 to 11 servings of bread, cereal, rice, or pasta each day; three to five servings of vegetables; two to four servings of fruit; two to three servings of milk, yogurt, or cheese; and two to three servings of meat, poultry, fish, dry beans, eggs, and nuts; in addition to eating fats, oils, and sweets sparingly. The guideline was memorable, Ratner said, though people had a hard time remembering the recommended numbers of servings of each food group. However, it was less actionable, because people were not sure about the size of a serving and it was hard to keep track of servings over the course of a day.

In 2005 the U.S. Department of Agriculture released a new guideline called MyPyramid consisting of five food groups: grains, vegetables, fruits, milk, and meat and beans. People were asked to go to a website where they would enter their age, sex, and how much exercise they got in a typical day, and the website would produce individualized guidelines for how much people should eat, in ounces and cups, from each of the five groups.

Ratner and her colleague Jason Riis studied the memorability and actionability of the food pyramid and found it wanting in both dimensions. After studying the personalized guidelines in each of the five categories for as long as they wanted, people were asked immediately after to recall the five numbers they had just seen. Only 19 percent of participants correctly recalled the numbers in all five categories, and less than 1 percent correctly recalled the correct numbers in all five categories 1 month later. People also were unsure about the size of an ounce of food, and again they had difficulty tracking their consumption over the course of a day.

A Better Way

Porter Novelli, the public relations firm that helped to develop MyPyramid and the original food pyramid, was testing another message at about the time that the food pyramid was introduced: fill half your plate with fruits and vegetables at every meal. Ratner and Riis saw this as a much more memorable and actionable recommendation, and testing confirmed their hunch. Immediately after studying it, 85 percent of participants correctly recalled the guideline, and 62 percent recalled it one month later. The guideline was also actionable, since people can tell when roughly half of a plate is full of fruits and vegetables and they did not need to keep track of their consumption over the course of a day.

When a guideline is memorable and actionable, people are more motivated to follow the guideline, Ratner stated. For example, in a comparison of the MyPyramid and the half-plate recommendations, both dieters and nondieters demonstrated more interest in adhering to the latter, with a particular increase in motivation among dieters.

Ratner said that she was delighted to learn that the new Obama administration intended to revamp the government's messaging about nutrition. In 2011 a new guideline, ChooseMyPlate.gov (http://www.choosemyplate.gov), incorporated the half plate of fruits and vegetables into a much more schematic treatment of the five food groups. This revision "definitely has the potential to help people follow nutrition science," she said.

Extensions Beyond Nutrition

In general, Ratner listed four attributes that make a message memorable and actionable, no matter what its subject.

First, it needs to be simple. Examples of simple messages are "got milk?" "drop and roll," and "just do it." They are easy to remember and easy to follow.

Second, the message needs to be easy to visualize. For example, the "got milk?" message is associated with a simple and easy-to-remember advertising image.

Third, a message should specify when to engage in an action. For example, when people were asked to take vitamins each day, those given an action plan for taking the vitamins—such as doing so every morning at breakfast—were nearly twice as likely to do so than those without an action plan.

Fourth, a message should embed a trigger to take action. For example, the dining hall message "live the healthy way, eat five fruits and veggies a day" was not nearly as effective as the message "each and every dining hall tray needs five fruits and veggies a day," because the latter reminded people to eat fruits and vegetables—but only in dining halls that had trays.

What Should Be Simplified?

Besides being memorable and actionable, messages need to be motivational and plausible, said William Hallman, professor and chair in the Department of Human Ecology at Rutgers University, in his comments on Ratner's presentation. For example, people need to understand why eating more fruits and vegetables is important. The graphic does not convey that information, so the reasons for eating some foods rather than others would need to be learned.

Also, not everything should be simplified. Much nutrition advice is simplified—"dangerously so," Hallman said. "Sugar is death" is a simple message, but in purely biochemical terms it may be more accurate to say that *no* sugar is death. Messages can be simple, plausible, memorable, and actionable—"and just plain wrong."

What Should Be Actionable?

Many science communications are not actionable, added Borchelt. Indeed, science communicators often would prefer that scientific information not be dragged into a political arena where it can be used to justify action of one kind of another.

Where action is desired, constraint recognition often stymies action. People may recognize that climate change is a dire problem, but they may also believe that nothing they do will make a difference.

People often do not have enough information to determine whether they have the ability to take action, Borchelt said. Simple and actionable messages can help people choose between two evidence-based diets, but that is not the typical situation in science communication.

ENHANCED ACTIVE CHOICE: A NEW METHOD TO CHANGE BEHAVIOR

When Punam Anand Keller, the Charles Henry Jones Third Century Professor of Management at the Tuck School of Business at Dartmouth College, was a child, she had the job of offering drinks to people whom her father counseled at their home, but many declined her offer. Finally her father asked her what question she was asking, and she replied, "Do you want a drink?" "Aha!" said her father. "That's your problem! You gave them an alternative." As soon as she began asking whether people wanted something hot or cold, more people accepted her offer.

Choice architecture, or the way options are designed and offered, can have a big influence on the decisions people make, Keller said. In particular, choice architecture is a useful communication method to simplify trade-offs when decisions have to be made by the audience. For example, simply telling someone to do something is not a very effective prompt for problem recognition. People are told to lose weight, save energy, volunteer, save money, and drink responsibly, but few do all five. Communication fails when it does not connect a person with a problem, a question, a goal, or a dream. It also fails when people are not motivated to take action or make a judgment based on the information conveyed in the communication. In contrast, messages that prompt people to take action can be a very useful tool to increase the effectiveness of the communication.

Choice Architecture

Keller described four of the most common forms of choice architecture. The first is an opt-in approach, as when colloquium participants are told, "Check the box if you will attend the Sackler webinar on the Science of Communications." The second is an opt-out approach, which would use the question "Check the box if you will *not* attend the Sackler webinar on the Science of Communications."

The third option is an active choice, in which participants pick among options. For example: 1. "I will attend the Sackler webinar on the Science of Communications." 2. "I will *not* attend the Sackler webinar on the Science of Communications."

The fourth option is enhanced active choice, in which the choices specify the advantages of choosing that option. For example, colloquium participants might be asked the following:

1. I will attend the Sackler webinar because it is important for me to discover new communication ideas and share research with academics and practitioners.
2. I will *not* attend the Sackler webinar because other commitments prevent me from discovering new communication ideas and the opportunity to share research with academics and practitioners.

In creating enhanced active choices, those designing the choices obviously face the issue of how directive to be and whether the choices are effective, Keller observed. But it can be an effective tool in getting people to take beneficial actions. Enhanced active choice can be personal, motivating, and interactive, which can help engage someone who would otherwise be uninvolved.

Encouraging Behaviors

As an example, Keller cited an enhanced active choice involving flu shots for hospital employees. The enhanced active choice asked the employees to check one of two boxes:

1. I want to remind myself to get a flu shot.
2. I want a reminder to get a flu shot.

Providing employees with an enhanced active choice was 50 percent more effective than the active choice of asking them to check either "I don't want a reminder to get a flu shot" or "I want a reminder to get a flu shot."

Another example involved enrollment for automatic refills of prescription drugs. When people were presented with a button on a website that they could click to enroll, about 12.5 percent did. But when a second button was added that said, "I prefer to order my own refills," thereby forcing people to make a choice not to enroll, almost twice as many people enrolled. Comparable results are seen with mailed responses, Keller added, so the outcome is not dictated simply by forcing people to choose.

Finally, Keller described an experiment with voice recordings designed to convince people to get their medications through the mail. In an opt-in approach, the recording said:

> "Would you like to speak with someone about getting started with mail service? Please say 'yes' or 'no.'"

In the enhanced active choice, people were told:

> If you would rather pay more and continue making many trips to the pharmacy, say 1. If you're tired of paying more and making unneeded trips to the pharmacy, say 2.

The enhanced active choice was 66 percent more effective, and despite how directive the choices were, most respondents reported being satisfied with how the automated voice response portion of the call was handled.

Enhanced active choice can provide a greater sense of control and belonging, be simpler and more urgent, and convey a sense of trust and shared goals, Keller concluded. As such, it meets the objectives of self-enhancement and accuracy desired of all science communication and can help nudge people in a scientifically accepted direction.

The Ethics of Enhanced Active Choice

Enhanced active choice clearly raises ethical issues, said Hallman, in his comments on Keller's talk. It is designed to persuade, not just to educate, which raises the question of whose interests are being served. After all, the standard question "Do you want fries with that?" could be made into an enhanced active choice question, which is not necessarily in a customer's best interest. Hallman also asked for a third choice for any such question: 3. "Please treat me as an adult and stop patronizing me because I can make my own decisions."

That said, Hallman added, people who most need to understand complex information are often in the worst position to understand it. Simplifying and creating choices that make sense for individuals and for society can, in these circumstances, have many advantages.

Enhanced Active Choice to Serve Science

Like upstream engagement, enhanced active choice is a tool, noted Borchelt. For science communication, a relevant question is whether the enhancement can incorporate science. The enhancements are often dictated by people with a vested interest in promoting a behavior. Will these people use science to shape choices, or will they pursue a different agenda?

If science is used to create an enhanced active choice, a related question is whether that use will harm the credibility of science. A goal of science communication is to build trust in science and in scientists. Incorporating science into decision-making architectures could backfire if people feel they are being manipulated, Borchelt said.

LESSONS FOR SCIENCE COMMUNICATION FROM BUSINESS

Business communication, including marketing and public relations, differs from science communication, but it nevertheless can provide some valuable lessons to science communicators, said Davis Masten, former head of the design consulting company Cheskin, and Peter Zandan, global vice chair and worldwide research practice group leader at Hill+Knowlton Strategies, during a joint presentation on the second day of the colloquium.

"Science is moving like a freight train," said Masten. Though people may know little about it, science is a major part of their lives. Yet, today, science communication is being outmaneuvered, and this outmaneuvering is going to increase. As Zandan pointed out, businesses are now spending more than $1 trillion to get out their messages, with about half that amount devoted to targeted marketing designed to reach individuals. Zandan was challenged to come up with an estimate for how much is spent on science communications, but it is probably less than a billion dollars (excluding education)—so less than a thousandth as much. In fact, businesses are spending $9.5 billion a year just to research the effectiveness of their messages.

Furthermore, many of the techniques businesses use to reach audiences and to assess the effectiveness of their messages are derived from the social sciences, Zandan continued. Ironically, the work of the scientific community has been adapted by business more than it has been by the scientific community.

Science is still respected much more than other professions, largely because of its devotion to integrity and truth. If science communicators could use this respect to amplify their messages, they could have a much greater impact than they do today.

New Technologies and Social Media

As Masten said, within a few years, more than five billion people will have smartphones worldwide. The average city now has a billion sensors in it, and the number will be 10 billion by 2020. This amount of power and connectivity could make science relevant to the choices people make every day, and science communicators could help make that possibility a reality.

Social media also have changed the nature of engagement, said Zandan. About a million people read the print version of the *New York Times* each day, but Facebook, LinkedIn, and other social media sites involve hundreds of millions of people. Furthermore, their use of these platforms generates data that can be used to increase and channel that use. Businesses are already using these data to look at the return on investment for their messaging. As Zandan said, "The effectiveness that these platforms have provided is truly transforming communications for business."

With social media, communications are not as expensive as with the mass media. A video can go viral at very little cost. Intelligence, creativity, and social savvy may be needed to create a popular video, but the potential to do so is not limited to business. Moreover, with a billion people a day using social media, even a penny per day from each of them to support science communication would represent a large amount of money.

The Need for Transparency

Zandan said that the hottest word in business today is "transparency." The social media have helped businesses realize that they need to be truthful and socially responsible because deviations from truthfulness will be played out in public.

In addition, everything that moves in business is being measured. Business is focused on its return on investment, but it no longer looks at this return purely in financial terms. The return is related to all the goals that a business is seeking, which includes financial calculations but also broader measures of accountability.

Science communicators could benefit by applying the same emphasis on metrics and accountability in their work. Research on the effectiveness of messaging in business is about 1 percent, Zandan noted. For science communication, he suggested that 3 to 5 percent be allocated to communication research. Furthermore, businesses are available to collaborate on this research, drawing on the research they have adopted from academia.

The Need for Partnerships

Science needs to maintain its independence and objectivity, but collaborations with business may be a way for science to be heard and embraced by the public. Without such partnerships, science will become a smaller voice among the American public. As Masten said, the trillion dollars spent by businesses on messaging generates a lot of noise, whereas today all science has is a whisper.

Companies, governments, and nongovernmental organizations have different missions and objectives than do scientists, but they generally want to do what is right, said Zandan. They are receptive to scientific evidence and curious, because they need to listen to survive. "They need you just as much as you need them."

Partnerships can involve all sectors of society. For example, the Science and Entertainment Exchange at the National Academy of Sciences has led to more than 700 consultations between the scientific community and the people who make movies and television shows.

Another way to maintain trust and engagement is to reach out to the 20 million students enrolled in colleges and universities in the United States. In addition, museums receive almost one billion visits each year in the United States, providing another superb opportunity to engage the public in science.

Science serves the nation and the world, Zandan and Masten concluded. If science is called into question, a loss of trust could damage a national and international asset.

INFLUENCES OF SOCIAL MEDIA

For decades, social science research has reflected a dichotomy between mass media that broadcast to large undifferentiating audiences and interpersonal communication among people talking with each other, observed Duncan Watts, principal researcher at Microsoft Research. But this traditional dichotomy has been dissolving. The mass media are fragmenting into different channels and platforms, facilitated by a surge in recommendation engines. At the same time, individuals are being empowered by new technologies to grow audiences that are in some cases as large as those of traditional network television. This nearly continuous distribution of production has led to the new concept of "mass personal communication."

Even as technology and the media landscape have changed, the questions asked in the social sciences have remained remarkably stable. In the 1940s Harold Lasswell laid out the essential problem: "Who says what to whom, through which channel, and with what effect?" It is a straightforward question, said Watts, but answering it is extraordinarily difficult. Observing who says what to whom is hard to do at scale, and the difficulty is compounded by the multiplicity of channels. And measuring the effects of all this communication is even harder.

The Web 2.0 revolution may be bringing the answer within reach. To social scientists, social media platforms are like telescopes were to astronomers, said Watts. They are instruments that make the invisible visible and enable new kinds of science. Social scientists can now do the same kind of science that physicists and biologists do. In particular, Twitter is an almost ideal platform to address Lasswell's question. Everyone from the president of the United States to private individuals communicating with their friends is on Twitter. They have a well-defined attention graph, because the only reason to follow someone on Twitter is to hear what that person has to say, which is different from a site like Facebook. Social scientists can track the diffusion of information using tweets containing shortened URLs, which encompasses not everything on Twitter but a lot. And a restricted version of influence can be measured by looking at retweets, click-throughs, and conversations.

Elite Users on Twitter

Watts described a research project in which he and his colleagues looked at everything on Twitter over an 8-month period, which they distilled down from 5 billion tweets to 260 million that contained bit.ly URLs. They then used Twitter Lists, which users employ to filter their streams, to produce what are essentially crowd-sourced labels for users named in the lists. For example, if someone is listed in thousands and thousands of lists that have the word "celebrity" in the title, that person is probably a celebrity.

Using this technique, they identified four classes of "elite" users: celebrities, people in the media, people in other kinds of organizations, and bloggers. They then ranked users by the frequency of being listed and measured the flow of URLs from the top 5,000 users in each category to the mass of Twitter users. Of the total number of URLs on a random user's newsfeed, almost half come from one of these 20,000 people, which represents a tremendous concentration of attention on a network as distributed as Twitter.

With the exception of the other organizations category, the top users within each category pay more attention to other users in the category, a phenomenon known as homophily. This latter observations makes sense, said Watts, since organizations use Twitter not only to broadcast information but to hear what people are saying about them.

However, the picture is different for retweets. Celebrities and organizations do not retweet to each other much, but the media and especially bloggers retweet within their categories more than they do to the other categories.

Identifying Influencers

These data also can be used to analyze the flow of information through the network. Research done since the 1950s has demonstrated that trusted ties are more important than the mass media in determining individual opinions. But not all people are equally influential. A category of people called opinion leaders act as filters between the mass media and the masses. These opinion leaders absorb what is happening in the media, decide what is interesting, and pass that content to other people.

This effect can be quantified using Twitter data. Almost half of media-originating URLs are received from people who can be identified as opinion leaders. These opinion leaders consume more media content than average users, are more active on Twitter, and have more followers. They also are followers, in that they receive much of their information indirectly. In the past, opinion leaders have been treated as "influencers" who act as mini-broadcast stations, and businesses have devoted substantial effort to identifying these influencers to use them as conduits for marketing. As one observer wrote, "Influencers have become the 'holy grail' for today's marketers."

This is an appropriate metaphor, said Watts, because the point of the Holy Grail is that it is never found. For example, after a video has gone viral, people can always come up with explanations for its success. But these explanations have not helped marketers or anyone else predict which videos are going to become popular, and the same observation applies to influencers.

The Twitter data substantiate this conclusion. Watts and his colleagues examined 74 million cascades where people retweeted messages containing bit.ly URLs, looking at the attributes of both the retweeters and the content of the tweets. Among the tiny percentage of retweets that traveled more than one or two hops from their source, the only attributes that made a difference were past local influence and the number of followers, and even those factors have only a small explanatory effect. All the other attributes measured had no effect on predicting which tweets would be widely disseminated, including how many people someone is following, how much they tweet, and when they joined. Surprisingly, the content of the tweet also does not influence the amount of dissemination.

Given the randomness with which tweets are retweeted, individual influencers are essentially impossible to identify. Instead, larger numbers of people need to be targeted to reach influencers. The only possible exceptions are individuals sharing the two attributes that make a difference in retweeting, who tend to be well-known individuals such as the President and celebrities or organizations such as the Weather Channel. But even given the possible advantages of targeting these individuals, a campaign to disseminate a message generally cannot ignore the large number of "ordinary influencers" with few followers.

Computational Social Science

The attributes of Twitter users may explain only a small part of how extensively their tweets are retweeted, said Noshir Contractor, Jane S. and William J. White Professor of Behavioral Sciences at Northwestern University and one of two commentators for the session, but that is not necessarily a bad thing. A world in which the social sciences could explain a substantial part of human behavior might not be a very enjoyable world.

One thing that Watts's research demonstrates is the value of computational social science, Contractor said. Social science typically has relied on surveys, experiments, interviews, and ethnographies, but now it can test theories at scale using data provided by social media. However, it is important to remember that this research involves only Twitter, and social science should not be reduced to Twitterology.

The Chaotic Evolution of Technology

Human desires and needs, including the need to communicate, remain the same over time, said Xeni Jardin, the founding partner and co-editor of Boing Boing, but the systems used to communicate have been changing very rapidly in the past few decades. In addition, platforms like Twitter evolve over time because their designers cannot predict how humans will use them. "That chaotic effect is part of what keeps me fascinated with technology," Jardin said.

As an example of how social networks can influence people's lives, Jardin described her experience with breast cancer. She was diagnosed in December 2011 and completed her primary treatment in September 2012. She accidentally live-tweeted her diagnosis, because she thought that she was a young and aggressively health-conscious person who could never get cancer and that the experience of having a mammogram would interest her tens of thousands of followers. Once she tweeted her diagnosis, she remained active on Twitter and other social networks, including her blog, throughout her treatment. She also became aware of a group on Twitter called #BCSM, which stands for breast cancer social media. Though she never went to a real-life support group, she was very active with a group of breast cancer patients and care providers through #BCSM.

This experience was "one of the most formative experiences of my life," said Jardin. It was a revelation about how up-to-the-minute and critically important science information can be disseminated and shared in a meaningful way through Twitter, in some cases even before care providers get the information. This particular group and network would be a valuable case study for social scientists to examine the real-world consequences of social networks, Jardin said.

CHARTING SCIENCE CHATTER THROUGH SOCIAL MEDIA

According to Deb Roy, professor at MIT and chief media scientist at Twitter, television and Twitter are intersecting to create a new hybrid form of communication. This new hybrid is an audience-driven movement in which people have chosen Twitter as a natural way to talk to each other while they watch television. In turn, the combination is driving change both in the television industry and in Twitter.

Television delivers a synchronous experience to a large number of people, as do many other experiences. For example, everyone in a given area experiences a sunset at the same time, and this experience can be quite different when someone is with another person rather than being alone. Two people can look at a sunset together and talk about it, which can completely change the primary experience, Roy noted.

Twitter is public, so that tweets can flow freely within the network but also pop up on the front page of the *New York Times*. It also is a fast medium, where the chances of someone reading a tweet drop off rapidly with time. It can be live when people come together at a scheduled time to communicate. This enables it to function something like a soundtrack to a movie; Twitter exchanges often take the form of a social soundtrack around life in the moment.

Combining these two observations, Roy pointed out that the number of tweets containing the word "sunset" peaks at the same time as the sunset around the world. People see the sunset and tweet about it, making a solitary experience a shared experience.

TV Tweets

The same phenomenon occurs with television shows. During the presidential debates between Barack Obama and Mitt Romney, memorable phrases, such as "binders full of women," immediately spread through Twitter. "I didn't have to wait for the media pundits," said Roy. "My own network plus some people I never met before but who entered the conversation in the moment and created that dynamic social network around this event influenced how I encoded what just happened."

The same thing is happening with sports, drama shows, and even advertisements. Roy has been involved in research that tracks the content of tweets commenting on a particular television show or advertisement. The result is a graph containing detailed information about the cross influences of connected audiences and content. These tweets happen both immediately after an event occurs and for a period of time afterward, as people discuss what they saw and what others said about it. They demonstrate both the social amplification of an observation over time and the scale on which such amplification occurs. In fact, said Roy, the Nielsen company is incorporating tweets into a new way of rating the viewership of television shows in the United States.

Promoting Science Through Twitter

Media companies have started using this phenomenon as a way of reaching potential viewers. For example, ESPN could insert a small clip from a basketball game into tweets mentioning the game. Someone watching the game could therefore be enticed to tune in when they received the tweet. In this way, observed Roy, a mass media experience can be delivered in a targeted fashion to a particular audience using social media.

The same process could be used for science communication, said Roy. Both good science and bad science are mixed into television shows. Either kind of science message could provoke a conversation on Twitter that could help disseminate information about science. Furthermore, some mass events have prominent science components, such as earthquakes, which could be used to disseminate science information, such as how to prepare for natural disasters.

Providing a context that makes a science event relevant to an audience often transcends Twitter or any other single medium, Roy acknowledged. But Twitter has the potential to create connections that did not exist before. It can have a profound impact on how primary experiences are interpreted—whether the experience is a sunset, a television show, or science.

Other Hybrids

Other hybrid media with symbiotic relationships have existed in the past, said Contractor in his comments on Roy's presentation. For example, when television became widely prevalent, *TV Guide* became the country's most popular print magazine. Similarly, in the early days of the Internet, people who watched soap operas engaged in extensive conversations about the shows on Usenet sites. In fact, said Contractor, people who did not watch the shows got so caught up in the Usenet conversations that they began watching the soap operas being discussed.

He also observed that television broadcasts can be accompanied by not only second screens but third, fourth, and fifth screens. People with particular interests—such as science—could discuss one aspect of a show, while people with other interests discuss other aspects. This might offer a way to galvanize a segment of the public that is more interested in science than other parts of the public.

One big difference between social media today and the Usenet conversations, said Jardin, is that today's social platforms are privately owned spaces. Perhaps something is being lost by not having a public forum in which these conversations can occur.

WHAT PREDICTS WHICH SCIENTIFIC FINDINGS ARE WIDELY SHARED?

When Katherine Milkman, the James G. Campbell Assistant Professor of Operations and Information Management at the Wharton School of the University of Pennsylvania, was in graduate school, she became curious about the list of most widely e-mailed articles on the *New York Times* website. Why did some articles make it onto the list while others failed?

To satisfy her curiosity, she had a web crawler built that visited the paper's website every 15 minutes and compiled the locations and full text of all of the news articles on the site as well as whether they were on the most e-mailed list. Over about 3 months she collected data on more than 7,000 articles, which she and a coauthor then analyzed to determine what factors determine which articles make the most e-mailed list.

Hypotheses to Test

Several ideas from the social sciences informed the analysis. First, people care deeply about the impressions they make on others. This would suggest that people would be more likely to share interesting, surprising, useful, or positive news to increase their self-enhancement.

Another motive might be to increase social bonding through the sharing of news. In particular, sharing emotional experiences can bring people closer together, which would suggest that articles evoking strong emotions might be shared.

Sharing also could be a form of emotional regulation. Some stories can provoke activating emotions such as fear or awe that people might share with others as a way of making sense of those emotions. Other stories might produce deactivating emotions such as sadness that would cause people to withdraw into themselves. In this case, stories that produce deactivated emotions would reduce sharing.

Factors That Increase Sharing

The analysis showed, first, that the position of a story in the newspaper matters. A 1-standard-deviation increase in the time a story spends as the lead article on the *New York Times* homepage increased its likelihood of making the most e-mailed list by about 20 percent.

As hypothesized, more interesting, surprising, and useful articles were more likely to make the list. More emotional stories were also significantly more likely to make the list, particularly stories containing more activating emotions as opposed to stories containing deactivating emotions.

Translating the Findings to Science

To explore the question of how these findings translate to the sharing of science, Milkman and her colleague Jonah Berger gathered data reported for the first time at the colloquium. They asked approximately 4,000 authors of articles that were published in leading science and social science journals in the first half of 2013 if they would provide lay summaries of their scientific discoveries. About 20 percent agreed, resulting in 845 summaries of new scientific discoveries. They then recruited a separate panel of 8,000 nonscientists to rate a randomly selected scientific summary, and they averaged these ratings to create a measure of how likely an article is to be widely shared.

The results show that findings published in psychology journals are the most likely to be widely shared, followed by economics journals, sociology journals, and, finally, science journals. Digging deeper into the data, articles about business, psychology, other social sciences, and mathematics are most likely to be widely shared, while discoveries in chemistry, human services, biochemistry, genetics, and ecology are least likely to be shared.

They then used an automated linguistic classification software program to count the frequency with which people were mentioned in each summary. More frequent references to humans dramatically increase the likelihood of science being shared. "We care about science about people," said Milkman.

Surprisingly, after controlling for the disciplinary affiliation of an author, summaries written by women are significantly more likely to be shared than summaries penned by men, which was also true of articles written by women in the *New York Times*. But when men and women were coauthors of a scientific article and each wrote a summary of the findings, the summaries were equally compelling. So, women are choosing to work on more sharable topics than men, but they are no better than men at describing the same findings.

Comparing summaries written by coauthors of the same article revealed several key features that increased shareability. For example, one scientist wrote of a finding:

> We're trying to build new types of crystal by combining layers from different materials. We've previously shown these can have many applications in digital and analog electronics. In this work we were able to turn light into electricity with a high conversion rate using our new structures made from graphene and tungsten disulfide, both atomically thin layered crystals.

A coauthor's description had a much greater likelihood of being shared:

> We produced a device that, although atomically thin, can strongly absorb light and convert it to electricity in a very efficient way. For every 100 photons of light, 30 are converted to electricity, which is a value comparable to the best solar cells in the market.

The features that make a scientific summary more shareable are very similar to those at work in the *New York Times*, said Milkman. If a summary is more interesting or more likely to reflect positively on a sender, it is more likely to be shared. In addition, summaries that explain why a result is more useful or are more emotionally resonant dramatically increase the likelihood of sharing. Being more positive also has a significant though small effect on shareability.

Interestingly, men and racial minorities are more likely to say that they would pass along scientific summaries than women and Caucasians. These groups view a given summary as more interesting, emotion inducing, useful, and comprehensible, and in the case of minorities, more emotion inducing and likely to reflect positively on them if shared.

By choosing their words carefully, scientists can increase the likelihood that their discoveries will be widely shared, Milkman concluded. They should emphasize why their work is useful, rely on emotional language, emphasize the positive, and focus on what is interesting and surprising.

Making Science More Shareable

As Contractor pointed out in his comments on Milkman's presentation, the features that make a story shareable do not have much to do with scientific information. He also pointed out that the discussion of sharing has been unidirectional. However, the public also has knowledge, insights, and data that it can usefully share with scientists, and listening to this input can help scientists convey their own insights more effectively.

Jardin observed that different kinds of emotions can generate shareability, from outrage over human exploitation to affection for kittens. She also pointed out that independent blogs have been under intense pressure to pander to clickability, linkability, and shareability. Science faces the challenge of becoming more shareable while staying true to the point of the work and not pandering to the lowest common denominator.

On that note, Watts added during the discussion session that every scientific message is not simple, every lesson is not easily digestible, and every result is not intuitive. Making complicated scientific results bite sized, palatable, and competitive with all other media risks undermining the work. Messages need to be propagated without undermining the integrity of science.

SCIENCE NARRATIVES: MASS MEDIA AND ETHICAL CONSIDERATIONS

For communicating science to nonexperts, narratives can be appropriate and meaningful communication tools, said Michael F. Dahlstrom, assistant professor in the Greenlee School of Journalism and Communication at Iowa State University. However, narratives also raise ethical considerations about when and how to use them to communicate science.

A narrative is a causally linked temporal sequence of events involving specific, human-like characters, said Dahlstrom, adding "You might also call it telling a story." Narratives are processed differently than are evidence-based arguments. The latter are context free. A fact can be removed from an argument and it still maintains its meaning. But a section cannot be removed from a narrative without losing its meaning and ruining the narrative.

Evidence-based communication begins with abstractions that can be applied to predict or explain specific situations. Narratives have the opposite direction of generalization, starting with specifics from which abstractions can be surmised. In fact, people will generalize from a narrative even when it is not representative of reality.

These two paths are not created equally, said Dahlstrom. Narratives are a critical way people make sense of the world, understand cause and effect, and interpret why people act the way they do. They are recalled twice as well and read twice as fast as evidence-based content. Calls to increase the teaching of evidence-based communication and logical reasoning are well placed, but they should not be interpreted to mean that narratives are an inferior way of thinking. On the contrary, narrative thinking may have given humans an evolutionary advantage by enabling them to figure out what others are thinking and might do.

Narratives in Science

The controversy over vaccines and autism provides a stark example of differences in the two ways of thinking, Dahlstrom noted. No evidence exists to link vaccines and autism, but narrative accounts of children demonstrating signs of autism a month after getting a vaccine can be very powerful. As the deputy director of the National Immunization Program once stated, "This is like nothing I've ever seen before. . . . It's an era where it appears that science isn't enough." Rather than lacking trust in science, Dahlstrom noted much of the conflict is likely due to the narrative and evidence-based information being comprehended through different processing pathways.

Narratives also matter for science communication because many nonexperts get most of their science information from the media, and especially from television and the Internet. Most of this information takes the form of narrative-based stories, since narratives are a format that elicits attention among audiences. As Dahlstrom observed, the media try to focus on personal actions, fit events into a meaningful time frame, and personalize abstract concepts, which all lend themselves to narrative treatments.

Entertainment also uses narrative formats, and modern technologies have made entertainment ubiquitous. Someone may watch a half hour of news on television and then watch three hours of entertainment, with most of it in the form of narratives.

The Ethics of Using Narratives

Narratives are intrinsically persuasive, said Dahlstrom. They imply a normative assessment while neither stating nor defending their assumptions. They also reduce the formation of counterarguments by transporting a viewer, reader, or listener into the narrative. The more people are engaged, the more likely they are to accept what the narrative is telling them. Furthermore, fictional narratives result in similar levels of persuasion as do nonfictional narratives, which is one reason why the National Academies has established the Science and Entertainment Exchange to work with Hollywood film and television creators on fictional narratives.

These observations raise serious questions about the ethics of using narratives to communicate science. First, is the underlying purpose for using narrative improved comprehension or improved persuasion? Is the appropriate role of science communication to persuade an audience to accept views about science or to clarify understanding and engage a wider public in a more vigorous debate? These are completely different goals, Dahlstrom said, but a narrative can be used for either. A narrative can aim to persuade by emphasizing the preferred side of a science issue through characters that either agree with the preferred side or learn to do so through the narrative. Or a narrative can aim to increase comprehension by using events that explain all sides of a science issue and portray a character who is neutral to the issue or multiple characters who embody the different sides of an issue.

The second question concerns the appropriate levels of accuracy to maintain. Narratives can have multiple levels of accuracy. In some cases, accuracy may be relaxed for the larger purposes of communication. For example, a character's motivations or actions, the settings, situations, events, procedures, and time frames may be more or less accurate and realistic and may all be used to communicate science. This happens even in science classrooms—for example, in discussions of frictionless surfaces, which do not exist in everyday life.

The representativeness of science narratives is a related factor. An audience will generalize from a narrative. Therefore, should the example chosen in a narrative be representative of a broader issue, or is it acceptable to use an outlier on which to base a narrative? The vaccine controversy is an example. Depending on whether the goal is increased understanding or persuasion, narratives may not be representative of reality.

The third ethical question is whether narratives should be used at all. They may violate expectations of how people think scientists should communicate. Science may be so linked with evidence-based communication that the use of narrative by a scientist may diminish credibility. Yet other stakeholders will likely be using narrative within the debate, Dahlstrom pointed out. Indeed, it may be unethical not to use narrative and surrender the benefits of a communication technique to the nonexpert side of a science topic.

Where Science Could Benefit from Narratives

Finally, Dahlstrom offered three questions regarding the future use of narratives in science communications:

- Do narratives help or hinder the desire to build trust between science and nonexperts?
- How can narratives meet the science communication needs of new media audiences?
- Can narratives help communicate science beyond human scale?

Human perceptual systems experience a very thin ribbon of reality. But to reason coherently about climate change, for example, people need to think in terms that extend far beyond a human lifetime. Narratives could help bring experiences that are outside of human scale within the realm of comprehension and consideration.

Countering Narratives with Narratives

Marty Kaplan, Norman Lear Professor of Entertainment, Media and Society at the University of Southern California's Annenberg School for Communication and Journalism, pointed to another example of narrative-based promotion: direct-to-consumer drug marketing. Many companies are marketing drugs directly to consumers for things like depression, insomnia, and restless leg syndrome, accompanied by long lists of side effects. Many such advertisements are structured as narratives, where someone starts out troubled and ends up happy.

Scientific discourse competes with other narratives that are components of billion-dollar campaigns. Science needs to be involved in these campaigns, and it needs to counter narratives with narratives. As Kaplan said, "You know the expression 'don't bring a knife to a gun fight'? I submit, 'Don't bring a data set to a food fight.'"

The other discussant at the session, Melanie Green, assistant professor of psychology at the University of North Carolina, Chapel Hill, cited recent studies done in her laboratory on whether narratives or statistical information are more persuasive. Not surprisingly, people who use statistics are perceived as more competent, while people who use narratives are perceived as warmer. Other research suggests that the use of narratives can increase empathy for outgroups. To the extent that scientists are considered outgroups and want to be perceived as warmer, their use of narratives could increase the public's receptiveness to scientific messages.

Narratives are analogous to science in that they deal with cause and effect, and one approach to writing a journal article, Green said, is to make it a good story. Finding the story in a data set can be useful in communicating science to a broader audience as well.

Finally, she noted that narratives are perceived in a social context. They are directed not just at individuals but at the social groups within which people live.

HOW SCIENTISTS TALK TO ONE ANOTHER ABOUT THEIR SCIENCE—AND WHAT THE PUBLIC HEARS

How do scientists actually communicate with each other, asked Kevin Dunbar, professor of human development and quantitative methodology at the University of Maryland in College Park. In the past they have used letters, journal articles, books, and presentations, while today they also have access to e-mail, Facebook, Twitter, and other new communication platforms. But all of these media have been built by humans as aids to the way the human brain works.

Analogies and Brain Activity

Dunbar and his colleagues have been studying communication in laboratory meetings in the United States and in Italy to learn exactly how scientists interact in those settings. They audiotaped and videotaped the meetings and interviewed the scientists before and after the meetings. For physician-scientists, they also compared interactions with patients to interactions with other scientists.

One prominent finding was that scientists relied frequently on analogies. They used local analogies within domains to fix problems, regional analogies with nearby domains to generate hypotheses, and long-distance analogies across domains to conceive of explanations, with their goals dictating the kinds of analogies they use.

Dunbar and his colleagues also have been using brain scans to identify parts of the brain that are more active when making an analogy. For example, the farther an analogy is from the source concept, the more active is a particular part of the brain known as the front polar cortex.

Using Analogies Effectively

Dunbar also has been involved in brain-scanning experiments where subjects are given data that are consistent or inconsistent with a hypothesis. Data that are consistent with a hypothesis generate activity in particular parts of the brain, while data that are inconsistent do not.

The use of analogies makes a difference in laboratory interactions. Laboratories where members have similar backgrounds, such as a laboratory where everyone is a microbiologist, tend to have greater difficulty using analogies effectively. In contrast, laboratories where members have different backgrounds, such as a laboratory that combines physicists and chemists, can use analogies more readily to make discoveries.

Gender Analyses

Finally, Dunbar briefly mentioned that female scientists and male scientists do not differ in their use of analogical reasoning and social interactions. However, men were more likely to assume that they knew the cause of unexpected findings, whereas women were more likely to set out to determine the cause of such findings.

Using Stories to Communicate Science

Dunbar's observations are a powerful argument for interdisciplinarity in science, said Green. They also shed light on how narratives can be used to reach particular types of audiences. Nonscientists often say that they cannot deal with math or that physics is too hard, but they do not say that they cannot deal with stories. Both analogies and narratives can make science more accessible to such individuals. At the University of North Carolina, for example, a program called "Scientists with Stories" is training scientists in storytelling techniques to help them better communicate their science while also helping them look for the stories in their own research.

Green also cited the importance of giving undergraduates research experiences so that they can learn that science is messy and hypotheses are not always confirmed. Even if they do not become scientists themselves, they will know how the research process works, which could increase their trust in scientific findings.

Science as Narrative

Kaplan observed that Dunbar's conclusions demonstrate that doing science is a narrative. Research has dead ends, surprises, mistakes, serendipity, and adventure. Even the choice of a problem to study involves ambition, competition, personalities, glory, and rewards.

Yet the drama of science is obscured in scientific papers, which are reverse engineered so that the outcome looks inevitable. Scientific papers are written from the perspective of "first-person invisible," said Kaplan, with the process of science removed from the scientific results. Even though the process of science is a compelling story, scientists typically ignore that story in describing their work.

TALES TEENS TELL: INTERACTIVE MEDIA COMMUNICATIONS CAN IMPROVE ADOLESCENT HEALTH

Narrative communications have a unique power to promote understanding, and that understanding can improve decision making, said Julie Downs, director of the Center for Risk Perception and Communication in the Department of Social and Decision Sciences at Carnegie Mellon University. Narratives can capture and hold people's attention and provide the basis for a fuller understanding through coherent arguments, vivid imagery, and a foundation for new knowledge. They make people want to know what comes next, which means that people are more likely to get to the end of the message. People can acquire a general understanding from a narrative, even if they do not recall all the details. They can learn even when they do not realize that they are learning.

To translate science into narratives, theoretical models that can serve as guides are useful. In the health field, social cognition models of health are examples, though other models can also be used. These models do not provide specific content, but they broaden thinking and result in better communications than those created with no theoretical underpinning. To determine what content needs to be included, however, developers need to use a systematic investigation of what is known and understood by the target audience.

The Narrative Content

Narratives can take many different forms, some of which work better than others. The initial narrative is probably not going to be the best communication. As a result, early versions of a narrative need iterative testing with members of the target audience. Are they understanding the narrative the way they should? Are audiences interpreting the word choices in a way consistent with the narrative's objectives? To the extent that the narrative offers advice, how practical is that advice?

Pilot tests with a target audience need to encourage criticism so that the narrative can be refined and tested. The goal is a narrative that people understand in the proper way, that explains the science comprehensibly, and that urges action in the appropriate circumstances.

A Narrative Targeting Sexual Decisions

Downs used the example of a narrative that helps teens avoid pregnancy and sexually transmitted infection. The narrative was delivered through interactive video, which is an effective vehicle for audiences that may be skeptical and lack patience, which is the case for adolescents. Interactive video gives teens a feeling of agency and structure as they choose which way to go in the narrative. Teens also are used to nonlinear forms of media such as games or streaming videos.

The narrative was developed with expert input of what adolescents need to know to make good decisions about sexual behavior. Unlike much of the sexual education adolescents get, the narrative took a nonpersuasive approach. It sought to convey how infections are transmitted and how teens can reduce the chance of infection.

Teens are overwhelmed by what appears to be highly scripted behavior, said Downs. They adhere to behavioral scenarios that play out the same way every time, in the same way that people know what to do when they eat at a restaurant. Teenage girls describe going to a party, finding their way to a private room with a boy, and engaging in sexual activities. They do not see themselves as having much agency to act otherwise.

Teens also underappreciate relative risks and lack health knowledge. They have been taught in their sexual education classes that there is no such thing as safe sex, but they nevertheless will figure out ways to go right to the verge of what they have been taught not to do. They do not have a good understanding of what is high risk and what is low risk. They know about HIV infection, but have little understanding of how other sexually transmitted diseases are different and what implications that has for transmission or treatment.

The narrative builds on teens' highly scripted behaviors to make them comfortable with the story. It has characters who follow scripted paths that pause several times with opportunities for decisions, at which point the narrative stops and the viewers are asked what they want to see the character do next. One option is to continue along the scripted path, but other options would get the character off that path. The narrative also provides suggestions for how to take these alternate paths, some of which are cheeky and funny, others of which are direct or evasive. The videos then provide a 30-second cognitive rehearsal in which viewers can think about how to apply those suggestions in their own lives. "We can't force them to think—if only we could—but we can at least force them to wait," Downs said. "During this 30 seconds, we hope they give this some thought and apply it to their own lives."

The videos also try to foster a better appreciation of relative risk through the metaphor of a risk scale that goes up and down. They point out that some behaviors are riskier than others and how to reduce the risks. They also explain reproductive physiology and attack misconceptions about, for example, how infections are transmitted.

A 6-month randomized controlled trial involving 300 subjects found that this approach resulted in decreased risky sexual behaviors and decreased sexually transmitted infections. A wider field trial was under way at the time of the colloquium that includes follow-ups and greater use of clinical outcomes and health records.

Taking Readers Out of a Narrative

A particularly intriguing aspect of this project, said Green, is its use of formative research to figure out what information people have and what information they need. That is a key step with these types of interventions that should be emphasized.

Green also played devil's advocate with regard to the use of interactive videos. Despite the time and technology that goes into creating them, largely the same experience can come from reading a book. Communicators need to think about when interactive technologies are helpful and when it is better to stick with low-tech options. The nature of the audience is one factor in making this decision. Another is the psychological process a message is designed to evoke.

Narratives can transport a reader into another world, Green noted. Readers become immersed in the storyline and identify with the characters. But if readers have to stop and make a decision, they can be taken out of the narrative. The benefits of making them take responsibility for the future course of the story must be weighed against the potential disruption to the narrative experience.

The Role of Edutainment

Entertainment education, or edutainment, is a field that has been studied for 50 years, observed Kaplan. It has a highly developed theoretical base, a set of best practices, and techniques to evaluate its impact. It is well known for the impact it has had in such areas as combating adult illiteracy, domestic violence, and public health problems.

More than a decade ago, the Centers for Disease Control and Prevention (CDC) recognized that people pay attention to health messages in entertainment even if they know the entertainment is fiction. In 2001 it formed the Hollywood, Health, and Society program, which has been run by the Annenberg School and functions essentially as the CDC's Hollywood office. The program has worked with hundreds of television shows to raise the profile of public health needs. For example, shortly before Atul Gawande's book *The Checklist Manifesto* came out, the program brokered a connection with the show "E.R." to have a life saved because a doctor was forced to use a checklist. The day after the show aired in New York, a conference of 150 surgeons watched the entire episode as a way to learn the value of checklists.

In "The Bold and the Beautiful," a show watched by 500 million people worldwide every day, the Hollywood, Health, and Society program was involved in a storyline in which one of the main characters confessed to his fiancé that he was HIV positive. The day that happened, the STD/HIV helpline spiked from 2,000 calls to 5,000 calls, a greater response than for every other public service announcement, campaign, and surgeon general's announcement.

SURGING SEAS: A COLLABORATION IN FIVE ACTS

Note: This transcript of the final presentation on the second day of the colloquium has been edited for length.

Act 1: All the Science That Fits to Print

Ben dials on his phone, and Gabrielle answers her phone.

GAB – Hello?

BEN – Gabrielle Wong-Parodi?

GAB – Yes?

BEN – This is Ben Strauss from Climate Central. One of your colleagues at Carnegie Mellon recommended you as an expert on communicating risk.

GAB – I'm flattered. What are you looking for?

BEN – I'd like some help sharing results from a large study I'm leading on U.S. vulnerability to sea level rise and coastal flooding.

GAB – Tell me more.

BEN – We're building an online tool to show our results. I think it's very important that we share these results with the communities that could be affected most, and with leaders, and it has to be a powerful tool. I have the feeling that people really don't get the danger that climate change poses, and that's a problem I want to help tackle.

GAB – That sounds good. An online tool could be really valuable.

BEN – We're generating a lot of data, and we don't want to dumb it down.

GAB – Alright. . . .

BEN – The first thing is, we're doing our analysis for every kind of place you can think of. We're analyzing every coastal state, every coastal county, every coastal city, town, even zip codes. We're looking at congressional districts, at state legislative districts, at federal and state agency districts, even at city council districts. We really want our work tailored for different audiences.

GAB – That's a big positive. It's well known that the more you can localize risk for specific audiences, the more you can command their attention.

BEN – We're doing our analysis, too, for a huge number of potential impacts. We're looking at housing, at population subgroups, infrastructure from power plants to airports to roads to rail; we're looking at critical facilities like hospitals or fire stations; we're looking at schools, churches, hazardous waste sites, military installations, parks, and more. Much, much more.

GAB – That's a lot.

BEN – And then we also overlay our results against a spatial index of social vulnerability, divided into three categories based on a standard deviation method, so we can show how the physical exposure intersects with communities' intrinsic response capacity. And we have results that assume levees, when present, are adequate in their protection, and a different set of results that assume that they are not adequate in their protection.

GAB – Alright. . . .

BEN – And, finally, most important of all, the time dimension. We made localized sea level projections at more than 50 water level stations, but also integrated them with local flood statistics to generate forecasts of flood risk, not just sea level. So we want to show sea level projections, annual flood risk projections, cumulative flood risk projections, plus how climate change multiplies risk, for each decade, and for 10 different water levels. We want to give users a choice of carbon emissions scenario, of sea level model, and of what percentile estimate to view. How can we *tweak* our presentation so it hits in the gut? So people really *get* it.

GAB – I see. I would like to help.

BEN – I'm so glad I called. Talk again soon?

Act 2: Education or Manipulation?

Gabrielle calls Ben.

GAB – Hi, Ben. There's something I want to talk about. You say you want people to *get* it. What does that mean?

BEN – That they understand the stakes. *Feel* them. And that the feeling fits the stakes and, ultimately, that action fits the stakes. I keep seeing surveys that show people ranking climate change low on their issue priority list. I know I'm a specialist, but that's tragic to me. I want people to have the right level of concern.

GAB – Listen, I sympathize. And I happen to agree on the threat level. But did you hear yourself? How could you know what the "right" level of concern is for someone?

BEN – Maybe not personally. But at least *professionally*. Risk priorities should be in the same order as risk ranks.

GAB – But people are not always rational, at least in the way that you would like them to be, or they may have different values and priorities. You can't just give them information and hope that it works. There's so much more going on here.

BEN – That's a big problem.

GAB – Albert Camus once said, "Fiction is the lie through which we tell the truth." What if the only way to get people to feel the "right" level of concern—to really prioritize climate change according to its *true* risk—is to leave them with a *false* impression?

BEN – That's a bigger problem. I need to stay true to the science. That's my foundation. That's who I am.

GAB – I needed to know that. I feel the same way. I'm not interested in spinning this.

BEN – So here's the deal: we work together for compelling communication—as powerful as we can make it—that leaves the audience with a proper understanding of the science.

GAB – Deal.

Act 3: Optimist or Pessimist?

Gabrielle calls Ben.

GAB – Hi Ben. Ben, you really have to slim down how you show projections. This isn't the control panel of the Starship Enterprise!

BEN – But I think it's important to give people a range of values and some choice depending on how much risk they feel they can tolerate. Besides, the answers depend on future carbon emissions—on top of all the model uncertainties.

GAB – Providing some simple, limited choices makes sense. Just not the control panel of the Starship Enterprise. I'd suggest giving three or four choices—say, on a spectrum from optimistic to pessimistic.

BEN – Optimistic to pessimistic. That seems like it could capture a lot of things—uncertainty in the models, the level of emissions, and maybe even luck. And maybe it would give more of a personal connection.

GAB – That's what I was thinking. But first I would like to test it.

BEN – Really? We're just talking about a couple of words here. I like your intuition.

GAB – You would be surprised at the power a couple of words can have—and how wrong intuition can be.

Several months later.

BEN – Well?

GAB – Wow. Some surprises. If there's one thing I've learned in my experience, it's that data often trump intuition. "Pessimistic" made people think that the situation was very bad, that it was a worst-case scenario, as I expected. However, it also made them think that the situation was hopeless, that nothing can be done about sea level rise. Listen to what one subject said: "All of our coastal communities and development are doomed." Then she added, "Seems as though fast rise can be dealt with, but 'pessimistic' makes me feel like nothing can be done."

BEN – Really.

GAB – And "optimistic" seems to make people think sea level rise isn't a problem at all. Listen to this: "That there's hope, it's okay that the sea level is rising, because it's rising slowly and we won't see any dramatic change soon. If we're optimistic about slow rise, then we don't really have to care."

BEN – So, we get rid of our bright idea, and keep the terms simple: "fast" or "slow."

GAB – That's what the data say.

Act 4: Near or Far?

Gabrielle calls Ben.

GAB – So the results are in for the first big experiment using our simple research tool. It turns out that concern is highest for the persons to whom we showed the year 2050 projections, not the 2020 or 2100 projections.

BEN – A rather balanced outcome in light of the conflicting forces we imagined.

GAB – People tend to care about the near term: the right here and right now. The farther off the problem is, the less they worry—all else equal.

BEN – But the farther off into the future that we project sea level rise, the more dangerous it becomes. So how do those opposite trends play against each other? How steep or accelerated does the sea level rise curve have to be before it evokes concern for the more distant future?

GAB – Good point. I think 2050 may be our sweet spot because it's far enough off to have a real sea level rise effect, but not much more than a 30-year mortgage away. It's a stretch, but maybe people can imagine themselves in 2050, or somebody they know. But 2100 just seems too far away, no matter how awful the scenarios are.

BEN – Okay. So we established in this experiment that projections for 2050 evoke more concern. But is that concern appropriate? How well did subjects actually understand the risk, and did it vary by treatment?

GAB – People *really* understood the 2050 numbers well. Equally as well as the 2020 figures. I was honestly quite surprised; I've rarely seen such good comprehension on a survey. However, there's a lot of confusion around the 2100 projections.

BEN – Well, that's nice to hear about the 2050 numbers! I think it provides some powerful guidance for this project. I do have a nagging doubt, though. We tested subjects using projections and flood statistics for New York City. How robust are the results going to be for different risk profiles from different places? Does each place have its own sweet spot year?

GAB – That's a really good point. I think I hear the sound of a new research proposal.

Act 5: Theory and Practice

Ben calls Gabrielle.

BEN – Gabrielle, we couldn't do everything you recommended, or even that our work together suggested. We had only so much money, time, and flexibility. We did try.

GAB – I understand.

BEN – You advised me to simplify. The website and pages should be broken into bite-size pieces with a clear order. We landed on the idea of breaking the tool into four page types to handle four main functions, that we call WHERE, WHEN, WHAT, and COMPARE. That's "mapping," "projections," "impacts analysis within communities," and "threats compared among communities."

GAB – I like it.

BEN – We broke the individual pages into sections, too. For example, with sea level rise and flooding projections in the San Francisco Bay Area, you can choose between the simple slow-through-fast sea-level rise scenarios we worked on labeling, how much carbon do you think we're going to put in the air, how lucky do you think we'll be, and how long do you think you'll live?

GAB – I like what you did. A few simple choices up front, like we discussed, but an "Advanced" tab, too, that's a bit harder to access.

BEN – Couldn't help myself. Users who want can choose specific sea level models, emissions scenarios, and low-range to high-range results given those parameters.

GAB – That's fair. Actually, it will be quite useful, I think, for some audiences. However, I'm a bit concerned that there will be too many options for some users and that it might be a bit confusing.

BEN – We did our best to explain, but I agree there's a risk.

GAB – We talked about focusing on one year—2050—but also about how using a simple sentence, versus a chart, would improve most people's understanding. As a scientist, it's easy to forget sometimes how hard it is for people to understand numbers.

BEN – You know I'm not completely sold on 2050 yet; and honestly I didn't really see how we could do it given the overall structure of the app. I take your points, though. We're planning to use those findings in other contexts—like one-page fact sheets and press releases. Here it just seemed important to provide richer data. But if you hover your mouse over one of the columns, you get a pop-up that explains the finding in a sentence or phrase.

GAB – That's terrific! I like the simplicity of the description, too.

BEN – Well, I didn't forget your lectures about writing for a junior high reading level. You chopped up too many of my sentences after your program evaluated them as written for philosophy majors.

GAB – Well, I hope people of all reading abilities can use this tool, and especially kids who are in junior high. They've got more at stake here than adults do.

BEN – That is too true. Look, I'm so grateful for your help, and can hardly believe that we're almost there. We're almost ready for our launch. Now comes the biggest test of all: how will people respond to the real thing, when it's really about their backyards.

GAB – I can't wait to see.

FINAL COMMENTS

In the final session of The Science of Science Communications II colloquium's second day, Dietram Scheufele, the John E. Ross Professor in Science Communication at the University of Wisconsin–Madison, identified four themes that struck him forcefully over the course of the day.

The first involves the role that scientists should play as arbiters of what is knowable. With controversies over vaccines, for example, scientists can determine the probabilities of certain things happening given particular levels of vaccination in the population. However, the policy implications of this information must be worked out through the democratic process, not in the scientific arena.

Second, the social and behavioral sciences have a fantastic new source of information in the data being generated by social networks. By making the invisible visible, these data provide scientists with information that they have never had before.

Third, collaborations involving business, scientists, and science communicators offer great potential, and not only in areas where science can help business sell more products. The field of science communication has much to learn from business that could be both unanticipated and extremely useful.

Finally, ethical issues play a surprisingly large role in science communication. What can be done is not always what should be done. Science communication needs to be held to a higher standard than most other forms of communication. That may put science at a political disadvantage, but failing to maintain high standards puts science at risk.

3

Creating Collaborations for Communication

In his welcome and orientation on the third day of The Science of Science Communication II colloquium, American Association for the Advancement of Science (AAAS) CEO Alan Leshner laid out the day's objectives. Each workshop participant was assigned to one of four breakout groups. These groups were charged with applying the lessons derived from the first 2 days of the colloquium to four pressing topics in science and science communications: climate change, evolution, obesity and nutrition, and nanotechnology. In particular, each group was asked to

1. Identify the challenges,
2. Segment the audiences,
3. Highlight the body of research,
4. Uncover the gaps,
5. Identify what is most important,
6. Spell out the contexts, and
7. Define and evaluate success.

Each of the breakout groups was to begin with presentations from content experts, communication scientists, and communication practitioners. Over the course of the day, the groups would then devise an action plan for science communication in each of their topic areas to be presented to colloquium participants in a final plenary session.

WORKING GROUP ON CLIMATE CHANGE

REPORT OF THE BREAKOUT GROUP ON CLIMATE CHANGE

Reporting during the final plenary session for the breakout group on climate change, Aaron Huertas, press secretary with the Union of Concerned Scientists, pointed to the challenge of communicating the relevance of climate change to members of the general public. But many communicators reach professionals whose jobs are affected by climate change. Among these professionals are the "first responders" to climate change, such as civic planners, water managers, coastal planners, military strategists, and meteorologists. Many of these professionals have to take climate change into account in their jobs, and they increasingly will have to do so in the future. They also tend to be nonpartisan, which means that they can largely avoid the political polarization that has characterized the issue. If more of these first responders were accurately reflecting messages derived from science, they could help break through the stalemate that currently surrounds discussions of climate change. In addition, research on how these professionals are integrating climate change science into their jobs and communicating the results to stakeholders could provide key insights into how to respond to climate change.

Many of the assessments done by national and international organizations are driven by stakeholders who need and ask for particular types of information. These requests for information and the data generated by these requests could be studied by social scientists to improve the effectiveness of public communications about climate change. For example, what explicit and implicit messages is the public receiving? Does the message that people are dealing with climate change today breed complacency or fear?

The breakout group developed several proposed actions. One is to have institutions use their convening power to bring scientists together with the people who make decisions based on climate science. These decision makers may have few opportunities at either the local or national levels to talk with each other or with scientists about climate change. The resulting networks of communication could involve scientists more closely in the decisions being made and in the dissemination of information about those decisions.

Climate scientists also would benefit by hearing from the people who use the information they generate. They would learn more about which stakeholders are using their research, thus enhancing their ability to point out how their research is affecting society. They also could improve their toolkits for effectively communicating about climate science to different audiences and help professional communicators more accurately convey scientific information to the public.

Professional norms for scientists will need to change for them to engage in this work. Their institutions need to encourage and reward scientists for getting out of their laboratories. Science education at the undergraduate and graduate levels could more explicitly include training in science communications. The high relevance of climate science to society creates strong incentives for such changes.

One measure of success would be more public voices validating and endorsing climate science. Nonpartisan voices outside the scientific community could help define what climate change means for the public. Another measure of success would be more local coverage of the effects of climate change and of local responses to change, which is likely to be less polarized than coverage at the national level. A final measure of success would be greater public perception of the importance of the issue. Today, many members of the public rank climate change as a relatively low-level concern. If climate science were more widely disseminated and understood, the salience of the issue would increase, Huertas concluded.

DISCUSSION DURING THE BREAKOUT GROUP

Most Americans do not have enough time to learn about climate change in depth, said Anthony Leiserowitz, director of the Yale Project on Climate Change Communication, during the discussion session of the breakout group on climate change. But if it were possible to convey five simple ideas about climate change to everyone, Leiserowitz's proposed list would be the following:

1. It's real.
2. It's us.
3. It's bad.
4. There's hope.
5. Scientists agree.

The climate communications community has not done an adequate job of communicating these ideas, said Leiserowitz. Yet if the American public understood and accepted these key ideas, people would be able to make more informed decisions both now and in the future.

According to polls, the majority of Americans—63 percent as of April 2013—currently believe that global warming is happening. But only about half of Americans believe that global warming is caused mostly by human activities, while a third believe that global warming is caused mostly by natural changes in the environment. Critically, only 4 in 10 Americans understand that most scientists think global warming is happening, and only 13 percent recognize that "81 to 100 percent of climate scientists think that global warming is happening." Leiserowitz described this last fact as a "gateway" belief—the more people without strong ideological responses (which is most people) understand the degree of scientific agreement about global warming, the more they themselves believe it is happening, human caused, and a serious threat and the more they support taking action.

The levels of skepticism among the public about global warming are not an accident, Leiserowitz continued. They have been substantially affected by media stories that pit a climate scientist against someone contesting the science and by what he called "a massive disinformation campaign by vested interests who are perfectly happy with the status quo." This disinformation campaign has borrowed heavily from a similar campaign that sought to convince Americans that the medical profession had not reached a consensus that smoking harms human health.

Global Warming's Six Americas

Leiserowitz and his colleagues have identified "Six Americas" that each have very different responses to the issue of climate change (Leiserowitz et al., 2013). They are (with percentages of the American public as of April 2013 in parentheses)

- Alarmed (16 percent),
- Concerned (26 percent),
- Cautious (25 percent),
- Disengaged (5 percent),
- Doubtful (15 percent), and
- Dismissive (13 percent).

These groups form a spectrum from the people who have the highest belief in global warming, are most concerned, and are most motivated to take action, to people who have the lowest belief and are least concerned and motivated. On the opposite ends of the spectrum, the Dismissive are outnumbered by the Alarmed. Yet the Dismissive are relatively vocal and tend to dominate public discourse, often giving the false impression that their numbers are much larger. Also, the U.S. Congress has a higher percentage of Dismissives than the general public, partly because the underlying electoral structure of American politics is increasingly politically polarized, Leiserowitz said.

When asked to identify the one question that they would like to ask an expert on global warming, members of the six groups gave different answers. The Alarmed and Concerned want to know what individuals and societies can do to reduce global warming. The Cautious and Disengaged want to know what harm it will cause and why they should care. Many of the Doubtful and Dismissive, however, want to know how experts know that global warming is happening or is caused by humans—and on a deeper level, why they should trust the experts. Of concern, said Leiserowitz, is the increasingly heard question "is it too late?" among some of the Alarmed, which is potentially dangerous because this conclusion may disempower those who believe in the need for action.

The Six Americas need tailored engagement strategies, Leiserowitz concluded. They interpret the facts in accordance with what they already know, value, and feel. Knowledge is necessary but insufficient. Emotions, values, ideology, and broader social, political, and economic forces all play critical (and often more important) roles in shaping public understandings and the political will to take action.

Fluctuating Concern

Nick Pidgeon, professor of environmental psychology and director of the Understanding Risk Research Group at Cardiff University in Wales, noted that concern over global warming has fluctuated over the past quarter century, with a high in the United States in 2001, according to polling from Gallup. Even though concern today is somewhat lower than this, it could increase again.

Researchers have looked at the factors that influence concern over global warming. Concern about the economy can displace concerns about the environment. Public fatigue over climate change stories and misleading press accounts based on leaked e-mails also have contributed to a decline in concern. Political polarization is increasing the number and vociferousness of skeptics in both the United States and the United Kingdom. Climate scientists, especially in the United Kingdom since the sizable press controversy over leaked e-mails from scientists in late 2009, have asked themselves whether they have lost the trust of the public, though polling in both the United Kingdom and the United States indicates that the loss has not been as great as some have feared.

Pidgeon pointed to three key issues in communicating about climate change. The first involves strategies to communicate about risk in the face of attempts to engender uncertainty. Some aspects of climate change remain uncertain, noted Pidgeon, but these uncertainties do not undermine the five key messages mentioned by Leiserowitz. One way to separate areas of uncertainty from areas of consensus is to separate risk assessment and decision making. People continually make decisions in the face of uncertainty. The challenge for scientists is to incorporate uncertainty into the information provided to decision makers in useful ways.

The second issue Pidgeon identified involves the narratives that are constructed to reach different audiences. For example, his group has been doing research on public attitudes and values regarding changes in the U.K. energy system. They have identified widespread public values, including the need to reduce the use of finite resources and overall levels of energy use. In turn, these public values are connected to other values ranging from a desire for social justice to a desire for autonomy and choice. The question then becomes how to construct narratives that go beyond the science of climate change and engage these widely shared values.

The third and final issue involves whether scientists should remain in their laboratories or emerge to become science communicators. Are they more likely to retain public trust if they limit themselves to describing the state of the science, or is there room for more engaged advocacy? Today, no consensus exists within the scientific community on this issue.

The State of the Science

Ralph Cicerone, president of the National Academy of Sciences, reminded the breakout group that climate involves much more than just the earth's global average temperature. Climate includes the extremes and patterns of temperature and precipitation, the amount of ice in the sea and on land, the temperatures, currents, and chemistry of the oceans, and so on. Furthermore, each of these variables is linked to various societal needs such as agriculture, water flows, and infrastructure.

The climate also changes naturally over time, both in specific locations and worldwide, and these changes will continue. The geological record documents prolonged periods of hot and cold, droughts, sea level changes, and movements of plant and animal species.

What is different today is that multiple lines of evidence point to human-induced climate change above and beyond natural climate change. People who think that the Earth's biogeochemical system cannot be changed by humans are wrong, said Cicerone. Some effects of increased greenhouse gases in the atmosphere are immediate, while others have time lags. Ocean currents change slowly, and a glacier can take many years to melt. But the eventual changes, even if hard to predict in detail, are potentially large and disruptive. Again, these changes include not just averages but the extremes. What will be the consequences when events expected to occur once a century on average instead occur much more frequently? How will the frequency of large fires in wild places change? The risks posed by these kinds of disruptive events warrant consideration and action today, said Cicerone.

Mitigation and adaptation are both necessary. Using less fossil fuel will have multiple benefits. Increasing resilience to extreme events, whether occurring naturally or as a result of climate change, is scientifically justified. At the same time, the development of good strategies is needed in such areas as geoengineering as people begin to talk about intentionally intervening in Earth's climate.

Scientific understanding continues to develop, Cicerone concluded. Conclusions made 20 or 30 years ago are being revisited and refined. New questions will arise as others are answered. But important questions posed in the early days of climate change research have been resolved, and climate science will continue to progress.

Enlisting Trusted Sources on Climate Change

Not only are people too busy to learn much about climate change, said Joe Witte, a researcher at George Mason University's Center for Climate Change Communication, but they are "cognitive misers"—they generally are not interested in the details of a scientific conclusion. However, most are willing to follow the advice of someone they trust, just as they trust and follow the advice of doctors without knowing all the details of what a doctor is advising.

A potential source of trusted advice on climate change is the television weather forecaster, Witte observed. They work in the same community as viewers and have many of the same values. They may not have enough time to go into the details of climate change, but they can provide a broad picture. And far more viewers are watching the local news on any given day than are watching such outlets as Fox News.

According to surveys, more than half of television weather forecasters want to talk about climate change, and some have already done so with great success. They may only be able to give the subject 30 seconds, but even that amount of time can convey the five messages mentioned by Leiserowitz. They also can break a longer treatment of climate change into short sections that can bring viewers back for more information.

Witte recommended that scientists adopt a television forecaster in their communities to disseminate information about climate change. About 15,000 weather forecasters serve more than 200 major television markets in the United States. If just a single forecaster in each of those markets made climate change a priority, the public would be exposed to much more climate change science than they are today.

To reach out to television forecasters, Witte recommended that scientists go slowly and think about how best to get their attention. The highest priority in local news is relevance to potential viewers. He said, "News directors will always ask a reporter, 'Why is your story, which you want to take a crew and report on, important to the viewers?'" Scientists can use quotations, metaphors, word pictures, comparisons, and other "grabbers" to capture the attention of forecasters. One formula for how to make things stick in people's minds is captured by the acronym SUCCESS—simple, unexpected, credible, concrete, emotional, and story or stories. Vivid images of how the climate is changing or how some aspect of the Earth system is reacting to climate change can capture a forecaster's and the public's attention and build a scientific case.

Local news is basically a headline service, said Witte. "If Moses were to come down today and say, 'Hey, I have 10 commandments everybody.' The local news director would say, 'Give me the first two.'" But forecasters can refer viewers to the web for more information, which allows viewers to become more informed while also recognizing the importance of the issue. It also cross-promotes the website for the TV station, Witte noted, thereby pleasing the station's sales force.

Surveys of weather forecasters reveal that many are worried about devoting some of their airtime to climate change science or about their capacity to be reporters. In response to these concerns, organizations such as Climate Central and NASA are producing videos and bullet points to make it easier to get climate information into forecasts. In this way, television forecasts can become a form of informal learning comparable to what happens at museums or zoos, Witte said. Audience research is also becoming more sophisticated, so that the prior conceptions of the audiences served by a local media market soon will become better known to broadcast meteorologists. This will enable specific audiences to be targeted, from the doubtful who mistrust scientific information to the alarmed who want to know what they can do to make a difference.

Maintaining Credibility

During the discussion session, the group discussed whether scientists risk losing their credibility if they enter the policy arena. As one participant observed, scientists are on a spectrum in terms of how comfortable they are talking about policy issues. Some would prefer to remain in their laboratories; others are eager to enter the political fray. The important point is that opinions about climate change are tied up with politics and personal beliefs, and the most effective science communicators are those who are aware of those beliefs and present science in a way that will not offend a listener's values.

Another participant pointed out that the existence of a consensus within the scientific community on the occurrence of climate change is not a political issue and can be emphasized without taking an advocacy position. Scientists also can explore the social and ethical dimensions of decisions related to climate change, and they can detail what is known and what is unknown or uncertain in a scientific or political domain.

Workshop participants also discussed the tendency for information to flow from schoolchildren to their parents when students are taught about such issues as smoking, seatbelts, and environmental hazards. The Next Generation Science Standards include material on climate change, which provides an opportunity to reach students. In particular, as one participant pointed out, success stories in which individual and societal changes not only reduce carbon emissions but bring other benefits are especially effective in engaging students and their parents. Such stories counter the hopelessness some people feel, create social support for behavior change, and demonstrate that humans can change the planet in beneficial as well as harmful ways.

WORKING GROUP ON EVOLUTION

REPORT OF THE BREAKOUT GROUP ON EVOLUTION

Reporting during the final plenary session for the breakout group on evolution, Robert Pennock, professor at Michigan State University, and Ann Reid of the National Academies observed that a strong consensus on the importance of teaching evolution in K-12 schools and in colleges and universities has emerged in recent years. Major national reform initiatives, including the *Next Generation Science Standards* (NGSS Lead States, 2013), the AP Biology Standards (College Board, 2012), and the report *Vision and Change in Undergraduate Biology Education* (Bauerle et al., 2011), have identified evolution as one of a handful of central concepts in biology education. This consensus has created an unprecedented opportunity to improve students' understanding and acceptance of biological evolution.

This opportunity will be lost, Pennock said, unless investments are made in implementing the recommendations of these initiatives. Instructors at all levels need new materials, administrative support, and professional development to be able to teach evolution effectively. Inquiry-based approaches in particular can enable students to build a deeper understanding of not just evolutionary processes and patterns but of how scientists use evidence to support hypotheses and reach conclusions.

A particular need, said Reid, is to develop a narrative that would get across the core concepts of evolution that students should learn. This narrative, which would be developed collaboratively by content experts, communication experts, and teachers, should emphasize the practical and positive benefits that evolution has in everyday life, with examples drawn from medicine, agriculture, ecology, and other fields.

The results of communications research can optimize these initiatives. Important questions include how to reach teachers, students, parents, school board members, and others with information that can convey important concepts and lower resistance to the teaching of evolution. New evaluation metrics could refine both education and outreach. Testing of the core narrative's impacts on teachers, students, and communities would lead to an iterative process of improvement. For example, what are the best ways to teach evolution without threatening religion or the sense of human specialness?

DISCUSSION DURING THE BREAKOUT GROUP

During the whole-group discussion of the breakout session on evolution, Pennock noted that different audiences require different messages and means of communication. When he was testifying about evolution and the nature of science in the 2005 court case *Kitzmiller v. Dover Area School District*, a critical need was to explain evolution and the nature of science without relying on jargon. Rather than discussing "methodological naturalism," for example, as one would in a philosophy of science class, Pennock explained generally how scientific explanations must be restricted to the physical realm of law-bound cause-and-effect relationships with no appeal to untestable supernatural powers. When speaking to a general audience, on the other hand, he shows a cartoon from *American Scientist* that effectively makes the point, in a simple, humorous way, that miracles are not allowed in science.

Creationists are very good at conveying their antievolutionary messages, Pennock added. They often describe evolution as "just a theory," drawing on the common meaning of the word *theory* rather than the scientific meaning of the word. They regularly speak of evolutionary biologists as "Darwinists," knowing that much of their audience will associate the term *Darwinism* with ideology and atheism. They describe the teaching of creationism in science classes as a matter of academic freedom, and they appeal to popular opinion with such catchphrases as "teach the controversy," or "teaching the other side is only fair."

Scientists need to counter such framing with their own framing, said Pennock. For example, supporters of evolution should refer to "scientists" rather than "Darwinists," to "evolutionary biology" rather than "Darwinism," and to "evolutionary science" rather than "evolutionary theory." Similarly, academic freedom entails the responsibility to teach science and not religion in science classes, and the central issue in science education is not fairness but integrity. In this way, scientists can respond to creationists with a framing that shifts the terms of the debate while also incorporating the values of science.

Communicating with the general public can require a different set of messages. As was noted on the first day of the colloquium, many members of the public may not be swayed by the opinions of a judge. Furthermore, opinions can vary widely among the public. Polls show that approximately 4 in 10 Americans accept evolution, 4 in 10 reject it, and 2 in 10 are undecided. Winning public favor for the teaching of evolution often means speaking to this middle group in ways that can reach them, said Pennock. In that regard, polls of religious beliefs can be misleading. Polls often ask questions that force respondents into a limited number of categories. For example, they can set up a false choice between God and evolution, whereas many theological positions are more subtle. For example, theistic evolution posits that God created the mechanisms of evolution and then set those mechanisms into action, which is a position that intelligent design creationism explicitly rejects. Even within evangelical Christianity, a wide variety of views toward evolution exist. For this reason, scientists such as Francis Collins, who is director of the National Institutes of Health and also an evangelical Christian, can be particularly good spokespersons for more nuanced views.

A specific audience that Pennock discussed is college students. The BEACON Center for the Study of Evolution in Action at Michigan State University (http://beacon-center.org) uses the idea of evolution in action to engage both in basic evolutionary research and in education. Instead of focusing on how evolution happened in the past, the NSF-funded center uses an inquiry-based approach to let students investigate evolutionary processes in real time, such as by using digital evolution (http://avida-ed.msu.edu). Students test hypotheses to learn how evolution works in such areas as medicine, agriculture, and engineering, using evolution to design robotic control mechanisms, for example, or exploring why a new flu vaccine is needed each year. In the process, they learn about evolution through evidence and inquiry rather than relying on the authority of a lecturer. By observing evolution in action and learning how to formulate evolutionary hypotheses, they correct their own misconceptions in a scientific way and have opportunities to learn about the processes, nature, and values of science. Evaluation of the program has revealed not just an increase in understanding but an increase in the acceptance of evolution.

The bottom line, said Pennock, is that the messages evolutionary scientists convey about evolution need to reflect the values of the audience being addressed, and those messages need to support the values of science.

A Formula for Effective Public Communication

According to Edward Maibach, director of the Center for Climate Change Information at George Mason University, effective public communication boils down to the following formula: simple clear messages repeated often by a variety of trusted sources. However, many scientists have not been very good at following that formula. They tend to work on complicated subjects and to believe that simplifying their work shortchanges it. But Maibach quoted a friend to the effect that "finding simple, clear messages isn't dumbing down what you know. It's smartening up what you know so that other people can understand it as well as you."

Scientists also tend not to like to repeat themselves. They prefer to talk about what is new and on their mind today, not what has been on their mind in the past. Yet the public generally does not absorb messages unless those messages are repeated, Maibach insisted. The first time people are exposed to a message, they often do not even hear it, much less understand it. The human brain is not motivated to process some forms of information, and repetition helps break down this barrier.

Finally, effective communication requires that a message be delivered by a variety of trusted voices. Research suggests that scientists are trusted, but they are not well known. When members of the public have been asked by pollsters to name a single living scientist, two out of three could not name a single one, and about half name scientists who have been dead for many years. Scientists need to work harder to become known to the public, Maibach said, and the scientific community needs to support these efforts. "Carl Sagan was a towering figure in America because he was willing to . . . put in the time at it, and he was really good at it. We don't have a lot of Carl Sagans in America today."

Science communications need to reflect this formula of simple clear messages repeated often by a variety of trusted sources. In the book *Science, Evolution, and Creationism* (NAS/IOM, 2008), for example, the cover, the chapter headings, and the text all conveyed and repeated straightforward and understandable messages, because many people could only be expected to glance at such a book. This typically requires audience research during the development of a communication to determine whether simple, clear messages are being delivered and whether they are achieving the desired objectives.

There is no such thing as "the public," Maibach concluded. There are many publics, which requires that science communicators decide which public needs to be reached. Audience research can enable such decisions by providing information about the values of an audience, how a message interacts with those values, and whether a communication advances a mission to the greatest degree possible. Success is not guaranteed, but such an approach can maximize the return on investments in science communication.

Finding the Story in Information

Journalists tell stories, they don't tell information, observed Dan Vergano, senior writer-editor at *National Geographic*. When journalists ask questions of researchers, they are looking for stories as well as information. When scientists only provide information, a story can be difficult to find. But if they provide journalists with at least suggestions for a story, they are more likely to get across the messages they want to convey.

Newspaper and magazine stories about evolution tend to fall into three categories, said Vergano. The first involves straight science stories, such as how an animal evolved certain traits because of environmental changes. The second involves the evolution of humans, which tends to generate pushback from some readers. The third involves controversies over evolution, such as coverage of the Dover area school board case. Stories in this last category are often handled by legal reporters rather than science reporters, which means that scientists need to be as simple and clear as possible to be understood. As Vergano quipped, whatever their other qualities, reporters are often the students who failed algebra.

Journalism may not be as powerful in shaping public opinions as popular television shows or movies, but it still can reach large numbers of people. Journalists help shape the public agenda and what people talk about on a daily basis. They also are good at identifying leaders and effective communicators within the scientific community.

Journalists cannot assume that their readers know much about science, so they need to explain things in simple terms. Scientists may be frustrated when they read these explanations in a publication, but they need to take the time and effort to educate reporters so that they can get the story right. Some science reporters are well versed in evolutionary science, but many are not. For example, reporters, like many members of the public, may be confused about the distinction between evolution and the origin of life, which requires that scientists be very clear in explaining these topics.

Science education in the United States has its weaknesses, but so does theological education, Vergano pointed out. Many people are poorly informed about their denomination's position on evolution. The leaders of a religion may accept evolution, but the members of that religion may not know about that acceptance—partly because they do not hear about that position from their leaders. These leaders would be a valuable potential audience for scientists.

Finally, Vergano pointed out that journalists are not on scientists' side. Their job is to convey reality as best they can and to generate good stories quickly. In addition, just as there is no public, there are no media. There are good and bad reporters everywhere, and they work for a wide variety of media outlets. Scientists need to find reporters who are good and feed them good stories because they then will become more prominent within their fields. They should think about these encounters as a campaign rather than a single exchange. Journalism is in the middle of a difficult period as it adjusts to new means of communication and new emphases within the profession. These changes also have given scientists more power to influence journalists. They can publicly discuss and publicize stories, including examples of good and bad journalism, on blogs and other outlets. Scientists "have tremendous power now that [they] didn't have 20 or 30 years ago to stick a wrench in the gears of the people who are causing the problem," said Vergano.

Overcoming Misconceptions About Evolution

Many people believe that accepting the occurrence of evolution requires giving up belief in God. As Eugenie Scott, executive director of the National Center for Science Education, said, this is "an extraordinarily toxic view which is hurting science literacy in this country." It also is demonstrably wrong. Many different theological positions exist, including the idea that God used evolution to achieve his plan, which is mainstream Christian theology. However, for people to learn about the continuum of viewpoints that exists, they have to be willing to listen, which means gaining their trust.

Another common misconception about evolution, according to Scott, is that humans evolved from monkeys. But humans did not evolve from monkeys or from apes. Rather, humans and apes share a common ancestor, just as we have common ancestors earlier in time with monkeys, fish, and petunias. The idea of branching lineages through time is not well understood, said Scott. Even people who have a grasp of natural selection often do not understand the big picture of a branching tree of life.

In her presentations, Scott explains evolution as a three-part idea. The first part is descent with modification, or common ancestry. The second and third parts center on the processes of evolution and the patterns of evolution. Patterns, for example, might focus on the relationships between bears and dogs or between birds and crocodiles. These patterns do not necessarily depend on natural selection, so scientific disputes about the validity of a particular pattern say nothing about whether evolution occurred. When creationists use such disputes to argue that evolution did not take place, they are committing a category error by using a dispute in one area to criticize an idea drawn from another area.

Layers of Understanding

Scientific understanding typically consists of three concentric layers, Scott said. At the core are the well-established ideas of science, such as common ancestry, that are used to explain natural phenomena. Around the core are frontier areas of science where research is occurring; these areas are typically the ones covered by reporters. Finally, on the edges are fringe ideas such as intelligent design creationism. Scientists do not spend any time on these ideas because they typically violate a core idea of science.

One of the most well-known statements in evolutionary biology is that "nothing in biology makes sense except in the light of evolution." Evolutionary thinking indicates why biological systems have the characteristics they do. All land vertebrates have four limbs because they are descended from an aquatic vertebrate that had four fins. Humans have 46 chromosomes and chimpanzees have 48 because, in the time since our common ancestor, two chromosomes merged in the human lineage, as clearly revealed by the Human Genome Project.

Scott describes her public presentations and interviews as drive-by science. She delivers a message and then is gone, with little or no opportunity for follow-up. A teacher, in contrast, can spend months with students and build on previous lessons. But even a single exposure to an idea can open a door so that the next time someone hears an idea it will be more comprehensible. A one-time presentation or interview is also an opportunity for education so that reporters have a better understanding of what the controversy over evolution entails. This information may not get into the story, but it can improve not only the current story but future ones. "Lower your expectations," said Scott. "Things will get better."

The website of the National Center for Science Education (http://www.ncse.com) provides much more information on all of these subjects, Scott concluded, as well as tools and information needed to counter the efforts of creationists and others to introduce nonscientific ideas into science classes.

Fairness and Objectivity

During the general discussion, members of the breakout session debated the proper role of journalists. One point of view is that the job of a journalist is to report accurately on not just the various sides represented in a dispute but on the balance of evidence supporting those positions. But Vergano emphasized that working journalists tend to favor fairness rather than objectivity. Someone may have an incorrect position, but if that position is having an effect on a community, journalists are obliged to report what that person is saying. They also have an obligation to report on the evidence that supports or fails to support that position, but they cannot ignore a story.

College Students as an Audience

Jay Labov, senior adviser for education and communications at the National Research Council, pointed out that the public is already getting simple clear messages repeated often—but they are coming from a well-financed and well-organized antievolution movement. The scientific community tries to react to these messages, but it largely fails at being proactive.

Labov particularly emphasized the importance of college students as an audience. For many college students, introductory courses in science are in fact terminal courses. These courses offer the last opportunities for instructors to teach students about how science differs from nonscience and about how scientists arrive at conclusions.

WORKING GROUP ON OBESITY AND NUTRITION

REPORT OF THE BREAKOUT GROUP ON NUTRITION

Reporting during the final plenary session for the breakout group on nanotechnology, Christine Wallace, president of Catalysis LLC, pointed to a long history of well-meaning nutrition education and awareness programs. Yet understanding of how to create change in this area remains elusive.

Change needs to be driven by the community. For that reason, community-based programs are needed that can be replicated in other locations with local people adapting programs to local needs. Community-based programs can mobilize community resources around issues of nutrition and food insecurity, create incentives, remove barriers, incubate programs, and support them as they develop. Both formative and summative research can inform other models developing in other locations.

In addition, programs must be based on and guided by good science so that they can continuously improve. Multilevel programs are needed to address heterogeneous settings and to secure sustained societal change. Two-way communication is essential for programs to be relevant and respectful. And scientists and communicators need checks and balances to ensure honest analysis and reporting of research. Examples include peer review of press releases, registered trials of nutrition interventions, broad assessments of the impacts of interventions, and diverse research paradigms that extend beyond randomized controlled trials.

The breakout group proposed creating an education campaign focused on coherent positive messages around nutrition. The program would be consistent with social, behavioral, and health research; adaptable to diverse settings; and adapted in response to research. A particular focus would be low-income families, and particularly women. A broadly based nutrition campaign could convey skills and knowledge that can help parents keep their children healthy. As an example of a short, positive message that could be part of such a campaign, Wallace cited, "Show your love, shape their future." The effects of such a message would need to be tested, but it is both positive and empowering.

In particular, the working group called for regional collaborative consortia around the issue of nutrition. These consortia would encompass community leaders, policy makers, businesses, and universities. The goal would be to raise the profile of the multigenerational implications of nutrition and food security, where poor nutrition in one generation has an impact on future generations. The consortia could draw on a clearinghouse for best practices and grassroots programs that have been successful, evidence-based resources drawn from research, and resources for advocates working with legislatures and other decision makers. The regional consortia could, in turn, coalesce into a national awareness campaign.

DISCUSSION DURING THE BREAKOUT GROUP

Ticketing people without seatbelts and raising taxes and restrictions on cigarettes were relatively easy ways of changing consumer behavior, said Julie Downs, director of the Center for Risk Perception and Communication in the Department of Social and Decision Sciences at Carnegie Mellon University. But obesity is a more complex issue. "You have to eat," she said. "We can't just ban food." Nutrition messages have to help people decide what to eat every day, and there is a great deal of nuance in those decisions. For example, research on food labels shows that they do not lead to consumption of fewer calories. Nor have labels on restaurant menus resulted in lower consumption. Downs speculated that the amount of cognitive processing involved in interpreting calorie information is part of the reason labels are not more effective. People's decisions about food are more general than mathematical, she said. For example, the 2,000-calorie-a-day campaign, designed to help consumers use the information on food labels, in most cases only makes choices more confusing.

In laboratory studies, caloric information does help people consume less, she explained. But these results do not translate into real life where there are so many other competing stimuli.

Research Activism

According to Brian Wansink, the John S. Dyson Professor of Marketing at Cornell University, the first stage of communication and collaboration consists mostly of wishful thinking, where researchers hope that what they are doing will be communicated in some way but do not take direct action. In the second phase, researchers take some control over communication by helping to write press releases. And in the third phase, they look critically at the research they are doing and work to adapt it to the needs of the present by addressing questions that solve behavioral issues and are scalable in practice.

Wansink described an epiphany he had when a senior colleague pointed out that his research was fairly obscure and accused him of wasting time. Wansink protested that his results could change people's behavior, but his colleague asked how that was going to happen. "It hit me like a silver bullet," Wansink said. He began writing a press release every time he published research. But he soon realized that the press coverage generated by his releases were not changing behaviors. Instead, his research had to be designed to be compelling, memorable, and clear. Research developed with activism in mind has a behavior-related outcome variable and does not require expensive and difficult interventions. Wansink now does every study three times—once in the laboratory, once in the field, and a second time in the field to fix whatever did not work the first time.

As an example, he cited a study of whether lower lighting and quieter music in restaurants encouraged diners to stay longer and eat less. "The world's not really interested in someone who can point out another problem, but they are very interested in people who can point out solutions," he said. With a sharp headline and a well-written article, the results of that particular study—showing that lower light and quieter music may reduce the number of calories that diners consume—make a compelling package that the researchers marketed to restaurants.

Journalism, Science, and the Public

Kathleen Zelman, director of nutrition at WebMD, talked about the realities of journalism and what to expect when journalists report on emerging science. Media companies are businesses, she said. Although journalists do their best to report accurately, sensational headlines, pseudoscience, and sky-is-falling headlines often get more attention.

Journalists are exposed to bad science all the time, she said, and not every study that is worthy of attention makes it into the news. For example, if writers cannot get in touch with a researcher the day a story is due, they cannot wait until later.

Communicating nutrition science is further complicated by misinformation disseminated by nonscientists who are vocal with their opinions. "Everybody eats," Zelman said, "and therefore, everyone is an expert in nutrition." Anyone can have a blog or post information on social media, creating a confusing welter of information for consumers. Furthermore, while Americans may not always act on health information, they are obsessed with finding it, and they often take it from noncredentialed sources.

People are aware that obesity is a problem, she said, but many people want a magic bullet and are drawn to celebrity diets and strategies that promise quick results. The how-to is what they need to know. "That's where we can really empower people." Both journalists and scientists have to begin by paying closer attention to how they translate science, with the goal of fostering public understanding.

Unique Communication Issues Around Obesity

David Allison, Quetelet Endowed Professor of Public Health at the University of Alabama at Birmingham, explored some of the practices that create myths and presumptions about obesity and nutrition. Many obesity campaigns target emotions, painting the issue as humorous, scary, or shameful. But "emotion doesn't always help people think well," Allison observed. Campaigns on obesity have not improved the accuracy of information or people's ability to interpret it. Some studies use language that implies causation when correlation is the only certainty. The more people hear a result, the more likely they are to believe it, whether or not it is true. But increasing belief does not increase knowledge.

If the peer-reviewed literature is not accurate, mistakes made by the media are hardly surprising. "Journalists get it wrong quite often. But are we complicit?" he asked, pointing to a press release that clearly misstated the results of a study. He also cited examples of inflated headlines and blatantly misleading articles drawn from perfectly precise journal articles.

Researchers have a tendency to layer studies on top of each other, continuing to do research that confirms already proven results. For example, studies showing that eating breakfast helps prevent obesity were first published in the early 1990s. Further research has confirmed the association beyond a reasonable doubt, but the studies keep coming out. This consumes time and resources without adding new information.

"It's not enough to point fingers at the journalists and the general public," he said. Scientists need to do better. One thought is to encourage scientists to take responsibility for providing truthful information. Another step is moving toward meta-methods such as clinical trial registries and public data-sharing policies. Finally, Allison encouraged researchers to promote public understanding of the processes of science so that the public can ask questions and take a more active role.

Clear Messaging

Peer review means more to scientists than it does to consumers, Zelman explained. The reputation of the person speaking often garners more trust, which explains some of the cachet of celebrity diets. "What we need to do for consumers is help them trigger that lightbulb moment," she said. "Help them make those decisions. They don't have to be perfect." Telling people they can do the right thing 80 percent of the time and still make a positive change helps them feel that they can succeed. Positive messaging and reassurance also contribute to changing behavior. We eat to live, she said, but also for pleasure. Taking the enjoyment out of food is not a sustainable approach.

Clarity is a crucial factor in communication, Zelman said. Simple, memorable messages will stick with people. The rule of thumb in the media is that no one will remember more than three messages. Apps, which are readily accessible to anyone with a smartphone, can help, but they also have to be simple, and wording is important. At WebMD, for example, surveys have shown that visitors particularly like slideshows that mix text and images.

Zelman emphasized meeting consumers where they are rather than expecting them to change how they find and process information. "It takes a village," she said. "That's why we're all here."

Overcoming Complexity

The complexity of obesity and nutrition messaging is daunting, breakout group participants agreed during the discussion session. Change is needed on many levels simultaneously, with shared responsibility among journalists, policy makers, and scientists. Communication about nutrition is a multidisciplinary and multiorganizational challenge requiring teamwork and synchronized messaging across time, since everyone makes immediate decisions about what they eat today and longer-term decisions about food and agricultural policy.

Context is important, as one participant attendee pointed out. Class affects which foods people have access to, while the environment shapes their food choices. Enhanced access to calorie-dense foods, subsidies for ingredients such as corn, and cultural and social forces all play a role in individual decisions. For that reason, healthy eating needs to be a shared goal, at both the family and community levels. Legislators, business owners, and food manufacturers are all part of the solution.

One participant pointed to the difference between explaining the science to consumers and simply giving them actionable information, speculating that the two are not necessarily compatible. It might be possible to give the public general rules of thumb to help them judge what information is sound, but evidence-based decision making is something even journalists have trouble with, so it could be too much to expect from consumers. Also, changing one behavior does not necessarily leave others intact. Many people compensate for exercise or healthier eating by practicing unhealthier habits in another part of their life, requiring that nutrition science investigate how those behaviors balance out.

Many strategies for changing behavior are abandoned when people do not achieve the desired result, said Jo Anne Bennett of the New York Department of Health and Mental Hygiene, rather than being considered one piece in a multidimensional communication strategy. "It can't be done with a single message," she said. She also pointed out that most people do not fit the description of "average," making 2,000 calories a day a poor guide for the majority of consumers.

Attendees agreed that positive messages, emphasizing the beneficial outcomes of healthy weight over the negative effects of obesity, were more likely to encourage consumers. Once people perceive a program and feel that they have the ability to change it, they will seek out solutions.

WORKING GROUP ON NANOTECHNOLOGY

REPORT OF THE BREAKOUT GROUP ON NANOTECHNOLOGY

Reporting during the final plenary session for the breakout group on nanotechnology, Darcy Gentleman, manager of public policy communications at the American Chemical Society, noted that at least some members of the public think of frankenfoods, grey goo, and being poisoned when they think of nanotechnology. The challenge is confronting these negative perceptions with the promise the science offers.

The breakout group proposed creating a large-scale training program for scientists that would teach them how to engage in two-way communications with different audiences. This training program would build on a prominent characteristic of the scientific community: its social nature. Scientists are constantly talking with each other in laboratories, during meetings, and at conferences. If they could learn to communicate as effectively with other audiences as they do with each other, they could leverage their social skills both in person and online.

A training program would create a community of practice among scientists that could compare and evaluate how messages are framed, how they are presented, and how they are received. Individual scientists could experiment with various forms of communication to explore which approaches work best. Effective approaches then could be leveraged through communication technologies to reach much wider audiences. In this way, the community of practice could cumulatively improve science communications, building on current expertise and past efforts.

One objective of the training program would be to create and empower champions of science communication, including champions who look like the audiences they are addressing. Some scientists, such as Mayim Bialik on the television show "The Big Bang Theory" and Neil deGrasse Tyson of the Hayden Planetarium, have demonstrated that they can attract very large audiences. In the case of nanotechnology, scientists trained in these fields also would have an understanding of the risks and promise of the technology and could convey that information to nonscientists.

DISCUSSION DURING THE BREAKOUT GROUP

Nanotechnology has a fast growth rate and fast bench-to-bedside transitions, which complicates discussions of social, ethical, and legal issues, explained Dietram Scheufele, the John E. Ross Professor in Science Communication at the University of Wisconsin–Madison. The science is also highly complex. Even someone with a good grasp of chemistry may not appreciate the many intersections between nanotechnology and biotechnology, information technology, cognitive science, and other fields. The speed, complexity, and breadth of nanotechnology make it an example of "a classic, wicked problem," Scheufele said, marked by "high policy stakes and high uncertainties."

When explaining nanotechnology, a convenient approach is to fall back on the wonder of how scientists have learned to make changes at a molecular level. But other aspects of the science are important for nonscientists, and especially policy makers, to know. Understanding the cultural differences and attitudes around how nanotechnology is received is crucial, Scheufele said. Some people are excited about new types of materials, but creating products that do not occur in nature can generate hostility. Without pertinent information about those products, nonscientists may form false assumptions.

How people translate their judgments of risks into attitudes about a specific product can be critical. Making one application of nanotechnology, such as medicine, more prominent in people's minds can influence the translation process and shape how they receive and even seek out new information. "Some of our research showed a weird paradox," Scheufele pointed out. "If you talk about applications, people get more excited about the technologies so that you essentially build buy-ins. But they're less likely to then want to find out more about the technology and its potential risks. So you essentially create a citizen who cares less about what they should be caring about. [But] if you talk about the risks and describe that these are new materials with new properties, . . . [people] immediately get a knee-jerk negative response to the technology."

Nanotechnology is an example of a technology where the expert community and the public are disconnected, Scheufele said. For many other technologies, scientists tend to be more enthusiastic and less pessimistic than the general public. With nanotechnology, in contrast, nanoscientists tend to be more concerned than the public about the potential environmental and human health impacts.

Building partnerships will be an important component of communicating about nanotechnology, particularly between social scientists and bench scientists. Institutionalizing these collaborations will be a way to increase market and political success. However, success may consist of deciding not to pursue particular technology, Scheufele concluded.

Engaging Scientists in Public Communication

Scientists tend to have low levels of public engagement, said Elizabeth Corley, the Lincoln Professor of Public Policy, Ethics, and Emerging Technologies at Arizona State University, and many scientists do not engage in public communication at all. In a recent AAAS survey of scientists across all disciplines, 93 percent rarely or never wrote about their results in a blog, only 3 percent talked frequently to reporters about their research, and just 39 percent talked with nonscientists. However, when asked about their level of interest in engaging with the public, 97 percent said it was an important part of their work.

Why the disconnect? Corley suggested that particular barriers prevent scientists from engaging in public communication, including lack of support, lack of time, and the fact that many scientists are not given credit toward tenure or promotion for non-peer-reviewed publications. She also pointed out that many graduate programs do not teach communication skills. And even scientists who do feel confident in their ability to explain their work may be put off by the well-documented culture clash between journalists and scientists. For example, a 2008 survey of scientists found that 90 percent saw risk of incorrect quotation as a disincentive for talking to the media.

How scientists view the media is correlated with their level of public communication, said Corley. Seeing media coverage as credible and comprehensive encourages scientists to disseminate their results more actively.

In summarizing the challenges to good science communication, Corley focused on content and engagement. Lack of communication training in graduate school makes for poor skills later in life, preventing some scientists from effectively conveying the important points of their work and what their results could mean. Engagement is the second part of the equation. If scientists do not make public communication a priority, they will not have productive interactions with the media or with the general public. Addressing institutional and cultural barriers along with the negative perception of media attributes would go a long way toward improving the status quo.

Building Trust

Nanotechnology is no different than other sciences when it comes to communication, said Julia Moore, the former director of legislative and public affairs at NSF who now works with the emerging issues team at the Pew Charitable Trusts. Many areas of science involve policy stakes, cultural values, risk values, and variation in public perception. "In the end, the public asks the same questions about every technology, no matter what they perceive as the risks and benefits," she said. "Who are the winners and who are the losers, because there always will be winners and losers. What are the risks, because nothing is safe? And, most importantly, who gets to decide?"

Moore emphasized the difference between communicating to educate and communicating to influence policy. An in-depth knowledge of nanotechnology is not desirable and potentially not possible with nonscientists, she argued. What matters is what they do with their perceptions. The most important job of a communicator is therefore to build trust. Trust, once established, eases the path from the initial explanation of a technology to eventual policy decisions.

To build this trust, scientists need to consider how their work may function in the wider world. "If you haven't thought about good regulations that are appropriate to 21st century technologies," Moore said, "then there is no reason for the public, policy makers, or journalists to trust you at all."

Many exciting uses for nanotechnology are present in aerospace and medicine, but most consumers encounter the technology in cosmetics, personal devices, and food. The familiarity of these domains is an important tool for science communication.

Moore also lamented the lack of communication training for scientists, but she pointed to examples of scientists who do well without training. Overcoming the reluctance to communicate and the fear of a backlash from the academic world is imperative to build trust with the public, she said. Talking to the media is part of scientific work, and every researcher has a responsibility to gain confidence and competence at explaining not only what they do but the implications of their work.

Regulation and Progress in Nanotechnology

Paul Weiss, director of the California NanoSystems Institute and professor of chemistry and biochemistry at UCLA, spoke about the opportunity to broaden uptake of nanotechnology. As demonstrated in the United States, Europe, and China, regulation and structure are key factors allowing nanotech to develop. Within that framework, scientists have great potential to connect across disciplines, fostering communication and collaboration.

Weiss used semiconductors as an example of moving technology from the microscale to the nanoscale. The smallest semiconductor structures produced today match the synapse scale of the brain, creating great potential for nanoscientists to work with biologists and neuroscientists on better understanding brain function. Nanoscientists have pushed hard to promote this type of work, contributing to the foundation of the BRAIN (Brain Research Through Advancing Innovative Neurotechnologies) Initiative announced by President Obama in April 2013.

Nanotechnology generates different attitudes in different countries. In Europe, cosmetics that use nanotechnology are not well received by consumers, so companies try to avoid it. France now has a registry for any product containing nanomaterials, which Weiss called "something to keep an eye on." In Japan and Korea, products with nanocomponents are more coveted than the alternatives and are more expensive. Nanomaterials are more difficult to test than chemicals, because at least 100,000 now exist. Some risks are fairly well understood, while others need further exploration, he explained. Testing will allow for better regulation and potentially remove some of the fears and negative perceptions around nanotechnology in consumer markets.

Levels of Public Engagement

Audience and panel members discussed how best to navigate the assumptions that the public forms about nanotechnology and what level of education is most effective. Moore, as someone who works to influence policy and regulation, said that her goal is not to explain the science in minute detail but to build general knowledge and encourage the public to advocate for oversight. Efficiency matters when it comes to policy changes, she pointed out. Her experience with a focus group on nanotechnology in cosmetics showed that consumers were unaware of the presence of the technology and overall not happy with the lack of information and lack of regulation.

Others agreed that nanotechnology's transition from the laboratory to the market went largely unnoticed by consumers. Research has shown that the first product people experience can have a large effect on their attitudes toward nanotechnology. One audience member pointed out that if scientists do not engage the public and give them accurate information, others will, citing Michael Crichton's 2002 science fiction novel *Prey* and the subsequent article published by Prince Charles expressing concern about nanotechnology.

In some cases, speaking to the media before an article goes through peer review can be a good thing, particularly where there is controversy or where risks and benefits are not well known, Corley argued. But peer review is not always slow. The first decision at *ACS Nano* takes only 12 days, Weiss pointed out, and liaisons with media allow their content to reach a wide audience.

Breakout participants observed that the scientific community should not focus only on policy makers but should take charge of their message and how it reaches the public. Moore expressed disappointment that risk analysis and regulation are still not very advanced and that industry has failed in many cases to be upfront about the presence of nanotechnology in their products and what it means. The group discussed the speed with which development of nanotechnology has outstripped the capacity for testing, and the need to rebalance priorities. Another attendee encouraged scientists to focus not only on communicating risk but also on expressing the potential of nanotechnology. One example is Taxol, a popular anticancer drug, which has greatly reduced side effects when made with a protein nanoparticle.

More education for scientists in how best to communicate is crucial, several participants argued. That training will help researchers address the challenges of talking about their work with the public and with policy makers, which require specific sets of skills. Institutions need to provide media training so the majority of scientists can be comfortable talking about their work to different audiences.

CLOSING REMARKS

In his closing remarks at The Science of Science Communication II colloquium, AAAS CEO Alan Leshner began by observing that the motivation for public engagement needs to be empowerment, not manipulation. People care about things that affect them personally or locally. Scientists therefore need to find ways to make their research and their messages personally meaningful and adaptable in a local seeing.

The public also needs opportunities to ask scientists questions. People cannot be seen simply as passive receivers of scientific information. They need to interact with scientists to understand and use science effectively.

The professional norms of science need to change so that engagement with the public is rewarded. More and more young scientists are interested in interacting with the public. The scientific community needs to encourage and support these efforts.

Finally, very few scientists are naturals at interacting with the public. These skills need to be learned, which requires opportunities for scientists to receive training and resources in science communication.

References

Bauerle, C., A. DePass, D. Lynn, C. O'Connor, S. Singer, M. Withers, C. W. Anderson, S. Donovan, S. Drew, D. Ebert-May, L. Gross, S. G. Hoskins, J. Labov, D. Lopatto, W. McClatchey, P. Varma-Nelson, N. Pelaez, M. Poston, K. Tanner, D. Wessner, H. White, W. Wood, and D. Wubah. 2011. *Vision and Change in Undergraduate Biology Education: A Call to Action,* C. Brewer and D. Smith, eds. Washington, DC: American Association for the Advancement of Science. Available at http://visionandchange.org/files/2011/03/Revised-Vision-and-Change-Final-Report.pdf.

Casman, E. A., B. Fischhoff, C. Palmgren, M. J. Small, and F. Wu. 2000. An integrated risk model of a drinking-water–borne cryptosporidiosis outbreak. *Risk Analysis* 20(4):495-512. Available at http://dx.doi.org/10.1111/0272-4332.204047.

College Board. 2012. AP Biology Course and Exam Description Effective Fall 2012, rev. ed. Available at http://media.collegeboard.com/digitalServices/pdf/ap/IN120084785_BiologyCED_Effective_Fall_2012_Revised_lkd.pdf.

FAS (Federation of American Scientists). 2007. *National Intelligence Estimate: Prospects for Iraq's Future: A Challenging Road Ahead.* Available at http://www.fas.org/irp/dni/iraq020207.pdf.

FDA (Food and Drug Administration). 2013. Structured Approach to Benefit-Risk Assessment in Drug Regulatory Decision-Making. Draft PDUFA V Implementation Plan—Fiscal Years 2013-2017. Available at http://www.fda.gov/downloads/ForIndustry/UserFees/PrescriptionDrugUserFee/UCM329758.pdf.

Leiserowitz, A., E. Maibach, C. Roser-Renouf, G. Feinberg, and P. Howe. 2013. *Americans' Actions to Limit Global Warming in April 2013.* Climate Interpreter. Available at http://climateinterpreter.org/sites/default/files/resources/Leiserowitz%20et%20al.%202013%20-%20American%27s%20Actions%20to%20Limit%20Global%20Warming%20in%20April%202013.pdf.

Mohan, D., M. R. Rosengart, C. Farris, B. Fischhoff, D. C. Angus, and A. E. Barnato. 2012. Sources of non-compliance with clinical practice guidelines in trauma triage: A decision science study. *Implementation Science* 7:103. Available at http://www.implementation-science.com/content/7/1/103.

NAS/IOM (National Academy of Sciences and Institute of Medicine). 2008. *Science, Evolution, and Creationism.* Washington, DC: The National Academies Press. Available at http://www.nap.edu/catalog.php?record_id=11876.

NGSS Lead States. 2013. *Next Generation Science Standards: For States, By States.* Washington, DC: The National Academies Press. Available at http://www.nextgenscience.org.

Royal Academy of Engineering. 2004. *Nanoscience and Nanotechnologies: Opportunities and Uncertainties.* RS Policy Document 19/04. London: The Royal Society. Available at http://www.nanotec.org.uk/finalReport.htm.

Appendix A

Agenda

MONDAY | SEPTEMBER 23, 2013

The Sciences of Communication

8:30-8:45	Welcome Barbara Kline Pope, Executive Director for Communications, The National Academies Moderators: Baruch Fischhoff (Carnegie Mellon University) and Dietram Scheufele (University of Wisconsin–Madison)
8:45-10:00	***Lay Narratives and Epistemologies*** Doug Medin (Northwestern University) Discussion with Ann Bostrom (University of Washington) and Kevin Dunbar (University of Maryland)
10:00-10:30	Break
10:30-11:45	***Motivated Audiences: Belief and Attitude Formation About Science Topics*** Susan Fiske (Princeton University)

	Discussion with Craig Fox (University of California, Los Angeles) and Bill Hallman (Rutgers University)
11:45-1:00	Lunch
1:00-2:15	***Communicating Uncertainty*** Baruch Fischhoff (Carnegie Mellon University)
	Discussion with David Budescu (Fordham University) and Adam Finkel (University of Pennsylvania)
2:15-3:30	***Social Networks*** Noshir Contractor (Northwestern University)
	Discussion with Deb Roy (Massachusetts Institute of Technology and Twitter) and Katherine Milkman (University of Pennsylvania)
3:30-4:00	Break
4:00-5:15	***Science Communication as Political Communication*** Dietram Scheufele (University of Wisconsin, Madison)
	Discussion with Kathleen Hall Jamieson (University of Pennsylvania) and Patrick Sturgis (Southampton University)
5:15-5:30	Wrap-Up/Lessons Learned Baruch Fischhoff (Carnegie Mellon University)

TUESDAY | SEPTEMBER 24, 2013

Science in a Time of Controversy

8:30-8:45	Welcome Ralph Cicerone, President, National Academy of Sciences
8:45-9:30	Keynote Address ***Responding to the Attack on the Best Available Evidence*** Kathleen Hall Jamieson (University of Pennsylvania)
9:30-10:00	Break

Moderator: Cara Santa Maria (Pivot TV)

10:00-11:30 **Panel 1 • Individual and Social Perceptions of Science: Three Cases**

Geoengineering: Public Attitudes, Stakeholder Perspectives, and the Challenge of "Upstream" Engagement
Nick Pidgeon (Cardiff University)

What Good Is a Guideline That People Can't Remember?: The Benefits of Extreme Simplicity in Communicating Nutrition Science
Rebecca Ratner (University of Maryland)

Enhanced Active Choice: A New Method to Change Behavior
Punam Keller (Dartmouth College)

Discussants: Bill Hallman (Rutgers University) and Rick Borchelt (U.S. Department of Energy)

11:30-12:15 ***Lessons from Business***
Davis Masten (Quantified Self) and Peter Zandan (Hill+Knowlton Strategies)

12:15-1:30 Lunch

1:30-3:00 **Panel 2 • Influences of Social Networks**

The Science of Social Media
Duncan Watts (Microsoft)

Charting Science Chatter Through Social Media
Deb Roy (Massachusetts Institute of Technology and Twitter)

What Makes Online Content Viral?
Katherine Milkman (University of Pennsylvania)

Discussants: Noshir Contractor (Northwestern University) and Xeni Jardin (Boing Boing)

3:00-3:30 Break

3:30-5:00 **Panel 3 • Narratives in Science Communication Science**

Narratives: Mass Media and Ethical Considerations
Michael Dahlstrom (Iowa State University)

How Scientists Talk to One Another About Their Science—and What the Public Hears
Kevin Dunbar (University of Maryland)

Tales Teens Tell: Interactive Media Communications Can Improve Adolescent Health
Julie Downs (Carnegie Mellon University)

Discussants: Melanie Green (University of North Carolina, Chapel Hill) and Marty Kaplan (University of Southern California)

5:00-5:30 ***Science Collaboration: Surging Seas***
Gabrielle Wong-Parodi (Carnegie Mellon University) and Ben Strauss (Climate Central)

5:30-7:00 Reception, Great Hall

WEDNESDAY | SEPTEMBER 25, 2013

Creating Collaborations for Communication

8:30-9:00 Welcome and Orientation
Alan Leshner, CEO, American Association for the Advancement of Science

9:00-9:15 Break and move to workshop locations

9:15-12:15 Concurrent workshops

Climate Change Workshop

Science Content Expert: Ralph Cicerone (National Academy of Sciences)

Communication Researchers: Tony Leiserowitz (Yale University) and Nick Pidgeon (Cardiff University)
Communication Practitioner: Joe Witte (NBC, retired)
Facilitator: Lynn Litow Flayhart

Evolution Workshop

Science Content Expert: Eugenie Scott (National Center for Science Education)
Communication Researchers: Robert Pennock (Michigan State University) and Ed Maibach (George Mason University)
Communication Practitioner: Dan Vergano (National Geographic)
Facilitator: William Courville

Obesity/Nutrition Workshop

Science Content Expert: David Allison (University of Alabama at Birmingham)
Communication Researchers: Julie Downs (Carnegie Mellon University) and Brian Wansink (Cornell University)
Communication Practitioner: Kathleen Zelman (WebMD)
Facilitator: Ellen Harvey

Nanotechnology Workshop

Science Content Expert: Paul Weiss (University of California, Los Angeles)
Communication Researchers: Elizabeth Corley (Arizona State University) and Dietram Scheufele (University of Wisconsin, Madison)
Communication Practitioner: Julia Moore (Woodrow Wilson International Center for Scholars)
Facilitator: Richard Tanenbaum

12:15-1:30 Lunch

1:30-3:30 Continuation of concurrent workshops

3:30-4:00 Break and return to auditorium

4:00-5:00	Reports from the workshops Moderator: Alan Leshner (American Association for the Advancement of Science)
5:00–5:15	Closing remarks Alan Leshner (American Association for the Advancement of Science)

Appendix B

Speakers

David B. Allison is internationally renowned for his research on obesity and statistical genetics. He has won numerous awards, including the Alabama Academy of Science's Wright A. Gardner Award and the American Society of Nutrition's Dannon Institute Mentorship Award in 2013, and the Presidential Award for Excellence in Science, Mathematics, and Engineering Mentoring in 2006. He has authored more than 450 scientific publications, edited five books, and is the founding field chief editor of *Frontiers in Genetics*. In 2012 he was elected to the Institute of Medicine of the National Academies.

Dr. Allison is Distinguished Professor, Quetelet Endowed Professor of Public Health at the University of Alabama at Birmingham, where he also serves as associate dean for science, director of the Office of Energetics, and director of the NIH-funded Nutrition Obesity Research Center. He received his B.A. in psychology from Vassar College and his Ph.D. in clinical psychology from Hofstra University.

Dr. Allison has earned a reputation as being a stalwart advocate for high standards of evidence for obesity research and science in general.

Rick Borchelt is director of communications and public affairs for the U.S. Department of Energy's Office of Science. Mr. Borchelt has held numerous high-level communications positions, most recently as special assistant for public affairs in the Office of the Director at the National Cancer Institute. He served as executive communications director for the Pew-funded Genetics and Public Policy Center at Johns Hopkins University and as communications director of the U.S. Department of Agriculture's Research, Education, and Economics Mission Area. He also has directed media relations for the National Academy of Sciences, has acted as press secretary for the U.S. House of Representatives Science Committee, and was special assistant for public affairs in the Executive Office of the President during the Clinton administration. Mr. Borchelt has worked overseas as well. He spent time in Nairobi, Kenya, as executive speechwriter to the United Nations under the secretary general and executive director of the United Nations Environment Programme. Mr. Borchelt received a B.S. in biology from Southeast Missouri State University.

Ann Bostrom is a professor of public affairs with the Evans School faculty of the University of Washington. Her research focuses on risk perception, communication, and management, as well as environmental decision making and policy. She has authored or contributed to numerous publications in these areas.

Dr. Bostrom also serves on the editorial board for *Risk Analysis* and as an associate editor for the *Journal of Risk Research and Human and Ecological Risk Assessment*. She was the recipient of the 1997 Chauncey Starr award for a young risk analyst from the Society for Risk Analysis. She received the award for her work on mental models of hazardous processes. Dr. Bostrom holds a Ph.D. in public policy analysis from Carnegie Mellon University, an M.B.A. from Western Washington University, and a B.A. in English from the University of Washington.

David V. Budescu is the Anne Anastasi Professor of Psychometrics and Quantitative Psychology at Fordham University in New York. His research is in the areas of human judgment, individual and group decision making under uncertainty and with incomplete and vague information, and statistics for the behavioral and social sciences. He is associate editor of *Decision Analysis and Psychological Methods*, and member of the editorial boards of *Applied Psychological Measurement*; *Journal of Behavioral Decision Making*; *Journal of Mathematical Psychology*; and *Multivariate Behavioral Research*. He is past president of the Society for Judgment and Decision Making (2000-2001), fellow of the Association for Psychological Science, and an elected member of the Society of Multivariate Experimental Psychologists.

Ralph J. Cicerone is president of the National Academy of Sciences and chair of the National Research Council. His research—on atmospheric chemistry, the radiative forcing of climate change due to trace gases, and the sources of atmospheric methane, nitrous oxide, and methyl halide gases—has involved him in shaping science and environmental policy nationally and internationally and has been recognized internationally by memberships in scientific societies and awards, including the Bower Award and Prize for Achievement in Science (1999) of the Franklin Institute, the Macelwane Award (1979) and the Revelle Medal (2002) of the American Geophysical Union, and the Albert Einstein World Award in Science (2004).

Dr. Cicerone's previous affiliations are the University of Michigan; the Scripps Institution of Oceanography at the University of California, San Diego; the National Center for Atmospheric Research; and the University of California, Irvine, where he was the founding chair of the Department of Earth System Science, the dean of the School of Physical Sciences, and chancellor.

Noshir Contractor is the Jane S. and William J. White Professor of Behavioral Sciences at Northwestern University and director of the university's Science of Networks in Communities (SONIC) Research Group. He is investigating factors that lead to the formation, maintenance, and dissolution of dynamically linked social and knowledge networks in a wide variety of contexts, including business, translational science, and engineering communities of practice.

Dr. Contractor has published or presented more than 250 research papers dealing with communicating and organizing. His book, *Theories of Communication Networks* (co-authored with Professor Peter Monge and published by Oxford University Press), received the 2003 Book of the Year award from the Organizational Communication Division of the National Communication Association. Dr. Contractor received a bachelor's degree in tech electrical engineering from the Indian University of Technology and his M.A. and Ph.D. in communication from the University of Southern California.

Elizabeth Corley is the Lincoln Professor of Public Policy, Ethics, and Emerging Technologies and an associate professor in the School of Public Affairs at Arizona State University. Her research interests focus on technology policy and environmental policy. Her recent book, *Urban Environmental Policy Analysis* (with Heather E. Campbell), was published by M.E. Sharpe in 2012.

Dr. Corley currently serves as a co-principal investigator and team leader for the $12.7 million National Science Foundation–funded Center for Nanotechnology in Society at Arizona State University. She is also a member of the editorial board for three peer-reviewed journals: *Evaluation & Program Planning*, *Journal of Technology Transfer*, and *Research Evaluation*. Dr. Corley received three engineering degrees (B.S.C.E. in Civil Engineering, M.S.C.E. in Environmental Fluid Mechanics, and M.S. in Environmental Engineering) and a Ph.D. in Public Policy—all from the Georgia Institute of Technology.

William J. Courville is a leadership educator, coach, and consultant. He has more than 30 years of experience in business, counseling, leadership development, and coaching. Dr. Courville has coached executives from a wide range of organizations, including Fannie Mae, the World Bank, and Lockheed Martin. He also has worked with senior leaders around the world.

In addition to his involvement in the corporate sector, Dr. Courville teaches management and coaching at many institutions, including University of Maryland University College and Georgetown University. Dr. Courville has a B.A. in philosophy from Loyola University, a B.S. in business administration from Louisiana State University, an M.Ed. in counseling from Loyola University (New Orleans), and a Ph.D. from the University of Ottawa (Ontario, Canada). He also received his certificate of Leadership Coaching from Georgetown University and is a Professional Certified Coach (PCC) with the International Coach Federation.

Michael Dahlstrom is an assistant professor in the Greenlee School of Journalism and Communication at Iowa State University. His research focuses on the effects of narratives on perceptions of science and the biases inherent when attempting to perceive topics beyond the realm of human scale. He is also part of an interdisciplinary, 3-year National Science Foundation grant to explore the ethical issues surrounding communicating science to non-experts.

Dr. Dahlstrom's research has been published in leading journals in the communication field, such as *Communication Research*, *Media Psychology*, and *Science Communication*, and he currently sits on the board of the Communicating Science, Health, Environment and Risk Division of the Association for Education in Journalism and Mass Communication. He also received the Shakeshaft Master Teaching Award in 2013. Dr. Dahlstrom received a B.S. in biophysics, a B.A. in journalism, and an M.S. in biophysics from Iowa State University and his Ph.D. in journalism and mass communication from the University of Wisconsin–Madison.

Julie Downs is director of the Center for Risk Perception and Communication in the Department of Social and Decision Sciences at Carnegie Mellon University and associate research professor in that department. Her research focuses on how social influences affect decision making, and how people can make better decisions by understanding the nature of these influences. Her work encompasses many areas, including developing interventions to help girls make better decisions about their sexual behavior and devising strategies to help e-mail users identify fraudulent e-mails.

Dr. Downs has published extensively in psychological, public policy, and medical journals. She has given invited addresses at a number of distinguished venues, including the National Academy of Sciences, the Institute of Medicine, and the United Kingdom's Health and Safety Executive. Dr. Downs received her B.A. in psychology from the University of California, Berkeley, and her Ph.D. in social psychology from Princeton University.

Kevin Dunbar is professor of human development and quantitative methodology at the University of Maryland in College Park and the director of the Laboratory for Scientific Thinking, Reasoning, and Education: Genes, Brains, Minds, and Creativity. He also has maintained his professorship in psychology at the University of Toronto, Scarborough.

Over the past 25 years, Dr. Dunbar has been conducting research on the nature of the scientific mind. He investigates the ways that scientists work in the lab, and the informal teaching strategies that professors use in biology and immunology labs. Currently, he is using functional magnetic resonance imaging to understand how the human brain reasons and changes as a function of education. He has published widely in his field, and his work has also appeared in the popular press. Dr. Dunbar received his B.A. and M.A. in psychology from the National University of Ireland and his Ph.D. in psychology from the University of Toronto.

Adam M. Finkel is the executive director of the Penn Program on Regulation and a senior fellow at the University of Pennsylvania Law School. He is one of the nation's leading experts in the evolving field of risk assessment and cost-benefit analysis, with 25 years of experience improving methods of analysis and making risk-based decisions to protect workers and the general public from environmental hazards. He is the recipient of the David Rall Award from the American Public Health Association for "a career in advancing science in the service of public health protection." From 1995 to 2005 he was a senior executive at the U.S. Occupational Safety and Health Administration.

Dr. Finkel has published more than 60 articles on risk assessment and management in the scientific, economic, legal, and popular literature. He received his A.B. from Harvard College, his M.P.P. in public policy from the John F. Kennedy School of Government, and his Sc.D. in environmental health science from the Harvard School of Public Health.

Baruch Fischhoff has expertise in decision making and risk analysis. As the Howard Heinz University Professor in the Departments of Social and Decision Sciences and Engineering and Public Policy at Carnegie Mellon University, he works with students studying the decision sciences.

Dr. Fischhoff has many professional affiliations. He is a member of the Institute of Medicine of the National Academies and is a past president of the Society for Judgment and Decision Making and the Society for Risk Analysis. He also has been a member of the Eugene, Oregon, Commission on the Rights of Women, the Department of Homeland Security Science and Technology Advisory Committee, and the Environmental Protection Agency Scientific Advisory Board, where he chaired the Homeland Security Advisory Committee. A graduate of the Detroit public schools, Dr. Fischhoff holds a B.S. in mathematics and psychology from Wayne State University and an M.A. and Ph.D. in psychology from the Hebrew University of Jerusalem.

Susan T. Fiske is Eugene Higgins Professor of Psychology and Public Affairs at Princeton University. She investigates social cognition, especially cognitive stereotypes and emotional prejudices, at cultural, interpersonal, and neuroscientific levels. Author of more than 300 publications and winner of numerous scientific awards, she has most recently been elected to the National Academy of Sciences.

Sponsored by a Guggenheim, her 2011 Russell Sage Foundation book is *Envy Up, Scorn Down: How Status Divides Us*. Her forthcoming book is *The Human Brand: How We Relate to People, Products, and Companies* (2014). With Shelley Taylor, she wrote a foundational 1984 text: *Social Cognition* (2013,), and she has written *Social Beings: Core Motives in Social Psychology* (2014). She has lately edited *Beyond Common Sense: Psychological Science in the Courtroom* (2008), the *Handbook of Social Psychology* (2010), *Social Neuroscience* (2011), the *Sage Handbook of Social Cognition* (2012), and *Facing Social Class: How Societal Rank Influences Interaction* (2012). Currently an editor of *Annual Review of Psychology*, *Science*, and *Psychological Review*, she is also president of the Federation of Associations in Behavioral and Brain Sciences. Her graduate students arranged for her to win the University's Mentoring Award.

Lynn Litow Flayhart is an organizational development consultant with 30 years of experience in many areas, including leadership assessment, executive coaching, organizational assessment and development, culture assessment and change management, strategic and business planning, group facilitation, and team building. Her clients include Dell, Fannie Mae, the International Monetary Fund, and the National Academy of Sciences.

An adjunct faculty member of the Center for Creative Leadership and its affiliate, the National Leadership Institute, Ms. Flayhart is a published author and frequent speaker. She has consulted on executive development with the Columbia University Senior Executive Program. Ms. Flayhart also was a member of the World Association of Business.

Craig Fox's research cuts across many disciplines: management, psychology, law, economics, neuroscience, and basic science. He studies behavioral decision theory, how people make decisions under conditions of risk, uncertainty, and ambiguity. He has been published in the top journals in multiple fields and is the cofounder of the forthcoming journal *Behavioral Science and Policy*.

Dr. Fox is the Ho-Su Wu Term Chair in Management at UCLA Anderson School of Management. He holds a joint appointment as professor of psychology in the UCLA College of Arts and Letters and professor of medicine at the UCLA Geffen School of Medicine. Dr. Fox is the cofounder and codirector of the UCLA Interdisciplinary Group in Behavior Decision Making. He received his B.A. in economics and psychology from the University of California, Berkeley, and his M.A. and Ph.D. in psychology from Stanford University.

Melanie Green is assistant professor of psychology at the University of North Carolina, Chapel Hill. As a social psychologist, she looks at the power of narrative to change beliefs, including the effects of fictional stories on real-world attitudes. Her theory of "transportation into a narrative world" focuses on immersion into a story as a mechanism of narrative influence. Dr. Green has examined narrative persuasion in a variety of contexts, from health communication to social issues. She has also investigated the influence of technology (in particular, television and the Internet) on social capital, and the ways in which trust can develop in online relationships.

In addition to publishing extensively, Dr. Green also has been recognized for her teaching. She received her B.S. in psychology and literature from Eckerd College and her Ph.D. in social psychology from Ohio State University.

William K. Hallman is professor and chair of the Department of Human Ecology, and is a member of the graduate programs in psychology, nutritional sciences, and planning and public policy at Rutgers University. Dr. Hallman has served on several National Research Council committees related to food safety risks, and currently serves on the Risk Communication Advisory Committee of the U.S. Food and Drug Administration.

Dr. Hallman's research examines public perceptions of biotechnology and nanotechnology in food production, risk perception and communication related to food safety risks, foodborne illness outbreaks, and food recalls, and public understanding and use of health claims made for food products and dietary supplements. He earned his B.S. from Juniata College and his M.A. and Ph.D. in experimental psychology from the University of South Carolina.

Ellen Harvey is the coaching talent manager for the National Leadership Institute. She is a licensed psychologist with 20 years of experience consulting for public- and private-sector organizations on the issues of executive coaching, leadership training, organizational development, and team building. She has held adjunct positions in the University of Maryland's Executive MBA and Executive Master's of Information Technology programs as well as Marymount University's graduate psychology program.

Kathleen Hall Jamieson is the Elizabeth Ware Packard Professor of Communication at the Annenberg School for Communication and the Walter and Leonore Annenberg Director of the Annenberg Public Policy Center at the University of Pennsylvania. She helped create FactCheck.org and FlackCheck.org, two nonpartisan projects of the Annenberg Public Policy Center that monitor deception in U.S. politics.

Dr. Jamieson is a fellow of the American Academy of Arts and Sciences, the American Philosophical Society, the American Academy of Political and Social Science, and the International Communication Association. She has won university-wide teaching awards at each of the three universities where she has taught and political science or communication awards for four of her books. Dr. Jamieson received a B.A. in rhetoric and public address from Marquette University and her M.A. and Ph.D. in communications arts from the University of Wisconsin–Madison.

Xeni Jardin is on the cutting edge of new media. She is the founding partner and co-editor of the award-winning blog Boing Boing. The blog is an eclectic selection of stories related to science, politics, and business, as well as human interest features. Ms. Jardin is also executive producer and host of the Webby-honored Boing Boing video, which is available online and in-flight on Virgin America.

Ms. Jardin has contributed to a range of traditional news and media outlets, including National Public Radio, *Wired*, *The New York Times*, CNN, and MSNBC. A founding board member of the Freedom of the Press Foundation, Ms. Jardin is committed to transparency in journalism and its important role in exposing law breaking and corruption in government. In 2011 Ms. Jardin was diagnosed with breast cancer and has used her own experience to educate women about the disease and raise awareness.

Marty Kaplan's career has had a broad reach, encompassing government and politics, the entertainment industry, and journalism. He is the Norman Lear Professor of Entertainment, Media and Society at the University of Southern California's Annenberg School for Communication and Journalism. He also was associate dean of the USC Annenberg School for 10 years and is the founding director of the School's Norman Lear Center, which studies the impact of media and entertainment on society.

In the political arena, Dr. Kaplan served as chief speechwriter to Vice President Walter F. Mondale. He also worked at Walt Disney Studios, was a featured commentator for NPR, and has been blogging for *The Huffington Post* since its inception. Dr. Kaplan graduated from Harvard College with an A.B. in molecular biology and received an M.A. in English from Cambridge University. As a Danforth Foundation Fellow, he received a Ph.D. in modern thought and literature from Stanford University.

Punam Anand Keller is the Charles Henry Jones Third Century Professor of Management at the Tuck School of Business at Dartmouth College. Her research focus is designing and implementing communication programs. In 2009 Dr. Keller was tapped to serve as the marketing director for the Financial Literacy Center, an endeavor launched by the RAND Corporation, Dartmouth College, and the Wharton School of the University of Pennsylvania and funded by the Social Security Administration. In 2010 her health communication model was selected by the Centers for Disease Control and Prevention as the best tool to tailor health communications (healthcommworks.org).

Dr. Keller's approach is to empower the voice of the consumer and to understand employee barriers to using financial and health benefits. She is past president of the Association for Consumer Research, a member association that connects researchers in multiple fields. Dr. Keller received her B.A. in economics and statistics from Elphinstone College, Bombay University, her M.B.A. from Bajaj Institute of Management, Bombay University, and her Ph.D. in marketing from Northwestern University.

Anthony Leiserowitz is widely recognized as an expert on how Americans and the international community view global warming, including public perception of climate change. As director of the Yale Project on Climate Change Communication and as a research scientist, he investigates the psychological, cultural, political, and geographic factors that influence public environmental perception and behavior. His work has addressed values and attitudes at the local, state, and global levels.

Dr. Leiserowitz has served as a consultant to many institutions, including Harvard University, the United Nations Development Program, and the Global Roundtable on Climate Change at the Earth Institute (Columbia University). He earned a B.A. in international relations from Michigan State University and a Ph.D. in environmental science, studies, and policy from the University of Oregon.

Alan Leshner has been chief executive officer of the American Association for the Advancement of Science (AAAS) and executive publisher of the journal *Science* since 2001. Before AAAS, Dr. Leshner was director of the National Institute on Drug Abuse. He has published more than 150 papers for both the scientific and lay communities on many topics, including science and technology policy, science education, and public engagement with science.

Dr. Leshner is an elected fellow of AAAS and a member (and on the Governing Council) of the Institute of Medicine of the National Academies, as well as a member of the National Science Board. He received an undergraduate degree in psychology from Franklin and Marshall College and M.S. and Ph.D. degrees in physiological psychology from Rutgers University.

Edward Maibach joined the George Mason University faculty in 2007 to create the Center for Climate Change Communication. Trained in public health and communication, he has extensive experience as an academic researcher and a communication and social marketing practitioner in government, business, and the nonprofit sector. His research focuses on the broad question of how public engagement in climate change can be expanded and enhanced. He has written numerous peer-reviewed journal articles, and the book that he edited, *Designing Health Messages*, earned a distinguished book award from the National Communication Association.

Dr. Maibach is currently a principal investigator on several climate change education grants funded by the National Science Foundation, NASA, and the Robert Wood Johnson Foundation. He also serves on the National Climate Assessment Development and Advisory Committee. Dr. Maibach received a B.A. in social psychology from the University of California, San Diego, an M.P.H. from San Diego State University, and a Ph.D. in communication research from Stanford University.

Davis Masten For more than 30 years, Davis Masten built Cheskin, a successful design consulting practice. The company focused on youth culture, branding, trust, and product development. Over the years, Cheskin worked on more than 2,500 projects in innovation for retail environments, corporate positioning, and identity.

In 2007 Mr. Masten sold the business to take on a new challenge—using science to address the needs of our growing population. To realize this goal, Mr. Masten serves as the cochair of the President's Circle of the National Academies and chair of the advisory board of Quantified Self, a collaboration of people interested in self-tracking to gather and share knowledge.

As part of his ongoing commitment to young designers and scientists, Mr. Masten is the Distinguished Visiting Scholar at the Media X program at Stanford University. He received his B.A. in marketing and psychology from the University of Redlands.

Douglas Medin is the Louis W. Menk Professor of Psychology at Northwestern University, with a joint appointment in the School of Education and Social Policy. His research focuses on three areas: concepts and categorization, decision making, and cross-cultural studies. In a partnership with the American Indian Center of Chicago, the Menominee tribe of Wisconsin, and Northwestern University, Dr. Medin and his team are studying how culture affects knowledge and reasoning about the natural world. He also has conducted research on cognition and learning among both indigenous and majority culture populations.

Dr. Medin is a recipient of an American Psychological Association (APA) Presidential Citation and the APA Distinguished Scientific Contribution Award and is a member of the National Academy of Sciences and the National Academy of Education. Dr. Medin also was the recipient of a James McKeen Cattell Sabbatical Fellowship Award for the 2010-2011 academic year and in 2013 received the Association for Psychological Science William James Award for lifetime achievement. He received his B.A. from Moorhead State College and his M.A. and Ph.D. from the University of South Dakota.

Katherine Milkman is an assistant professor at the Wharton School at the University of Pennsylvania. Her research relies heavily on "big data" to document various ways in which individuals systematically deviate from making optimal choices. Her work has paid particular attention to the question of what factors produce self-control failures, such as exercising too little or eating too much junk food, and how to reduce the incidence of such failures.

Dr. Milkman has published extensively in leading social science journals, and her work has been covered in the popular press, including *The New York Times* and National Public Radio. In 2011 Dr. Milkman was recognized as one of the top 40 business school professors under 40 by Poets and Quants. She received her B.S. degree from Princeton University in operations research and financial engineering and her Ph.D. from Harvard University's joint program in computer science and business.

Julia Moore has had a long career in public policy, with a focus on international science, technology, and security issues. She is a senior scholar at the Woodrow Wilson International Center for Scholars, and senior officer, Emerging Issues in the Government Performance Group at The

Pew Charitable Trusts. Previously, she served as deputy director of the Wilson Center and Pew Project on Emerging Nanotechnologies. She also is a former senior advisor in the Office of International Science and Engineering and past director of legislative and public affairs at the National Science Foundation.

Ms. Moore has worked for numerous other organizations. She is a past executive director of Physicians for Social Responsibility and past deputy director of the Arms Control Association at the Carnegie Endowment for International Peace, as well as a former Dean and Virginia Rusk Fellow at Georgetown University's Institute for the Study of Diplomacy. Ms. Moore received her B.S. from Georgetown University's School of Foreign Service.

Robert T. Pennock is a professor at Michigan State University, where he is on the faculty of Lyman Briggs College and the Departments of Philosophy and Computer Science. His research involves both experimental and philosophical questions that relate to evolutionary biology and cognitive science, such as the evolution of altruism, complexity, and intelligence. A national leader in evolution education, he was an expert witness in the historic *Kitzmiller v. Dover Area School Board* Intelligent Design creationism case. He also leads the Scientific Virtues Project.

Dr. Pennock is a fellow of the American Association for the Advancement of Science, and has served on the AAAS Committee on the Public Understanding of Science and Technology and the National Academy of Sciences *Science, Evolution, and Creationism* authoring committee. Dr. Pennock received his B.A. in philosophy and biology from Earlham College and his Ph.D. in history and the philosophy of science from the University of Pittsburgh.

Nick Pidgeon is professor of environmental psychology and director of the Understanding Risk Research Group at Cardiff University in Wales. Dr. Pidgeon's research centers on risk, risk perception, and risk communication, with a focus on the interface of social psychology, environmental sciences, and science and technology studies. Currently, he is studying public responses to energy technologies and climate change risks.

Dr. Pidgeon has led many policy-oriented projects related to public response to environmental risk for United Kingdom government departments, research councils, and the Royal Society. He also has published widely in his field. Dr. Pidgeon is currently a member of the UK Department for Energy and Climate Change's Science Advisory Group and a theme leader for the Climate Change Consortium for Wales. He received a B.A. in mathematics and psychology from the University of Keele and a Ph.D. in psychology from the University of Bristol.

Barbara Kline Pope is executive director for Communications at the National Academies and executive director of the National Academies Press. She is responsible for innovative and dynamic programs designed to bring science, engineering, and medicine to public audiences, including the Science & Entertainment Exchange and the Science and Engineering Ambassador Program. The Sackler Colloquium series on the Science of Science Communication is part of her portfolio of communication activities and reflects her deep interest in social and behavioral research.

Ms. Pope combines her responsibilities for communicating to diverse audiences with supervision of the National Academies' publishing programs, through which its lists of both scholarly and trade books have been available on the Web free to read since 1995. She is the president-elect of the Association of American University Presses, serves on the corporate advisory board of the marketing department at the University of Maryland, and also on the management board of MIT Press. Ms. Pope holds an M.S. from the University of Maryland.

Rebecca Ratner is professor of marketing in the Robert H. Smith School of Business at the University of Maryland. Her research focuses on factors underlying suboptimal decision making, with an emphasis on variety seeking, consumer memory and motivation, and the influence of social norms. She has been published in many decision-making journals, including the *Journal of Consumer Research* and the *Journal of Personality and Social Psychology*. She currently serves as associate editor for the *Journal of Consumer Research* and *Journal of Marketing Research*.

Dr. Ratner also has been recognized for her teaching. She received the Allen J. Krowe Award for Teaching Excellence from the Robert H. Smith School of Business in 2010. She earned a B.A. in psychology from Williams College and her M.A. and Ph.D. in social psychology from Princeton University.

Deb Roy is a tenured professor at the Massachusetts Institute of Technology (MIT) and is chief media scientist at Twitter. He conducts research at the MIT Media Lab on language, games, and social dynamics at the intersection of artificial intelligence and cognitive psychology. In 2008 he cofounded and was the founding CEO of Bluefin Labs, a social TV analytics company, which *MIT Technology Review* named as one of the 50 most innovative companies of 2012. Bluefin was acquired by Twitter in 2013. An author of more than 100 academic papers in machine learning, cognitive modeling, and human-machine interaction, his TED talk, Birth of a Word, has been viewed more than 2.5 million times.

A native of Canada, Dr. Roy received a B.S. in applied science (computer engineering) from the University of Waterloo and a Ph.D. in media arts and sciences from MIT.

Cara Santa Maria has a passion for science education, making it her mission to ensure that the fascination she feels for the natural world is shared by as many people as possible. Currently, Ms. Santa Maria cohosts *Take Part Live* on Pivot TV. She has also cohosted the Weather Channel's original series *Hacking the Planet* and *Young Turks* Network. Previously, she was the senior science correspondent for *The Huffington Post*. She coproduced and hosted a science talk show pilot for HBO and cohosted an episode of *StarTalk* Radio with Neil deGrasse Tyson.

Ms. Santa Maria is a North Texas native who currently lives in Los Angeles. Prior to moving to the West Coast, she taught biology and psychology courses to high school students and college undergraduates. Her published research has spanned topics from clinical psychological assessment to the neuropsychology of blindness. She received a B.S. in psychology and philosophy and an M.S. in neurobiology from the University of North Texas.

Barbara Schaal's career as a leading evolutionary biologist began with a youthful fascination with plants. Currently the Mary-Dell Chilton Distinguished Professor and dean of the Faculty of Arts and Sciences at Washington University in St. Louis, she is known for her work on the genetics of plant species. In particular, she has been recognized for her studies that use DNA sequences to understand evolutionary processes, such as gene flow, geographical differentiation, and the domestication of crop species. In her research, Dr. Schaal often collaborates with students and researchers from the Missouri Botanical Gardens.

In 2005 Dr. Schaal became the first woman elected to the vice presidency of the National Academy of Sciences. Since 2009 she has served on the President's Council of Advisors on Science and Technology. Born in Berlin, Germany, Dr. Schaal grew up in Chicago and graduated from the University of Illinois, Chicago, with a B.S. in biology. She earned her Ph.D. from Yale University.

Dietram Scheufele has received much recognition for his work in mass communication. He is one of only two scholars to win both early career awards in the discipline: the Young Scholar Award for outstanding early career research from the International Communication Association and the Hillier Krieghbaum Under-40 Award for outstanding achievement in teaching, research, and public service from the Association for Education in Journalism and Mass Communication.

Dr. Scheufele is the John E. Ross Professor in Science Communication at the University of Wisconsin–Madison and co-principal investigator of the Center for Nanotechnology in Society at Arizona State University. His current research program focuses on public opinion dynamics surrounding controversial science, with an emphasis on the interplay among media, policy makers, and lay audiences. He also consults with many organizations, including PBS, the World Health Organization, and the World Bank. Dr. Scheufele received his B.A. from the Johannes Gutenberg-Universität Mainz and his M.A. and Ph.D. from the University of Wisconsin–Madison.

Eugenie Scott has spent most of her career fighting for science. As the founding executive director of the National Center for Science Education (NCSE), she has been a researcher and an activist in the creationism/evolution controversy for more than 25 years. She is well versed in many aspects of the controversy, including educational, legal, scientific, religious, and social issues. Recently, the NCSE has taken on climate change, refuting the claims of those who question its validity and supporting teachers and other citizens in their efforts to keep good science in the classroom.

Dr. Scott has received national recognition for her NCSE activities, including awards from scientific and educational societies and humanist groups. She is the author of *Evolution vs. Creationism* and co-editor, with Glenn Branch, of *Not in Our Classrooms: Why Intelligent Design Is Wrong for Our Schools*. Dr. Scott received her B.S. and M.S. from the University of Wisconsin–Milwaukee and her Ph.D. in physical anthropology from the University of Missouri. She also holds nine honorary degrees.

Ben Strauss serves as chief operating officer and director of the Program on Sea Level Rise at Climate Central. In the latter capacity, he has published multiple scientific papers, testified before the U.S. Senate, and authored the *Surging Seas* report, which led to the development of SurgingSeas.org, a coastal flood risk tool. This tool has been widely covered in the popular press and has led to multiple appearances on television and radio outlets.

Previously, Dr. Strauss was a founding board member of grist.org, an environmental website, and helped launch the Environmental Leadership Program. He also co-organized the Campus Earth Summit and authored a report on college environmental education and practices for the Nathan Cummings Foundation. Dr. Strauss received a B.A. in biology from Yale University, an M.S. in zoology from the University of Washington, and a Ph.D. in ecology and evolutionary biology from Princeton University.

Patrick Sturgis is professor of research methodology at the University of Southampton in the United Kingdom, director of the Economic and Social Research Council National Centre for Research Methods, and president of the European Survey Research Association. His research interests are in the areas of survey methodology, statistical modeling, public opinion and political behavior, and public understanding of science and technology. He also has worked on the processes of inter- and intragenerational social mobility and public attitudes toward science and technology, as well as engagement with these subjects.

Currently, Dr. Sturgis is principal investigator of the Wellcome Monitor Study. He has a B.A. in psychology from the University of Liverpool and an M.S. and a Ph.D. in social psychology from the London School of Economics.

Richard Tanenbaum is president of Behavior Consultants and Global Learning Partners, LLC, specializing in executive coaching and leadership/organization development. His experience as a clinical psychologist, nonprofit executive, civilian working with the military, and actor enable him to work well with diverse clients from administrative staff to senior executives in any industry.

Dr. Tanenbaum's clients include the American Institute of Architects, Freddie Mac, the INOVA Health System, the World Bank, the International Monetary Fund, the Inter-American Development Bank, the National Institutes of Health, the United States Naval Academy, the World Bank, and all branches of the military.

Dr. Tanenbaum earned his Ph.D. and master's degree in clinical psychology from Virginia Commonwealth University. He received his bachelor's degree in psychology from the University of Pennsylvania. He is a certified coach for the Center for Creative Leadership and serves on the faculty of the National Leadership Institute at the University of Maryland and the Uniformed Services University. A licensed psychologist, Dr. Tanenbaum is certified in a wide spectrum of psychological and leadership assessment instruments.

Dan Vergano is a senior writer-editor at *National Geographic*. He spent the previous 14 years as the science writer for *USA Today*. He has written widely on areas that include climate change, evolution, and archaeology. Before entering journalism, Mr. Vergano worked as a space policy analyst and aerospace engineer.

Mr. Vergano has received numerous awards for his work. He won the 2011 Gene Stuart Award from the Society for American Archeology for a story and video series on Maya archeology, and the 2006 David Perlman Award for Excellence in Science Journalism from the American Geophysical Union for a *USA Today* cover story on climate change. Mr. Vergano was also a 2007-2008 Nieman Fellow at Harvard University, where he concentrated on the intersection of science and politics. He received a B.S. in aerospace engineering from the Pennsylvania State University and an M.A. in science, technology, and public policy from George Washington University.

Brian Wansink is the John Dyson Professor of Consumer Behavior at Cornell University, where he directs the Cornell Food and Brand Lab. Dr. Wansink's award-winning academic research on eating behavior, behavioral economics, and behavior change has been published in the world's top marketing, medical, and nutrition journals. He also is the author of a book for the popular press, the 2006 best-selling *Mindless Eating: Why We Eat More Than We Think.*

His work has had practical applications. It contributed to the introduction of smaller "100 calorie" packages to prevent overeating and removing 500 million calories from restaurants each year via Unilever's Seductive Nutrition program. From 2007 to 2009, Dr. Wansink served as executive director of U.S. Department of Agriculture's Center for Nutrition Policy and Promotion, the federal agency charged with developing the 2010 Dietary Guidelines and promoting the Food Guide Pyramid. Dr. Wansink received a B.S. from Wayne State College, an M.S. from Drake University, and a Ph.D. in consumer behavior from Stanford University.

Duncan Watts is a principal researcher at Microsoft Research and a founding member of the MSR-NYC lab. From 2000 to 2007, he was a professor of sociology at Columbia University, and from 2007 to 2012 he was a principal research scientist at Yahoo! Research, where he directed the human social dynamics group. He has also served on the external faculty of the Santa Fe Institute and was recently named an A. D. White Professor at Large at Cornell University.

Dr. Watts' research on social networks and collective dynamics has appeared in a wide range of journals, including *Nature*, *Science*, and *American Journal of Sociology*. He is also the author of three books, most recently *Everything Is Obvious Once You Know the Answer: How Common Sense Fails Us*. Dr. Watts holds a B.Sc. in physics from the Australian Defense Force Academy, from which he also received his officer's commission in the Royal Australian Navy, and a Ph.D. in theoretical and applied mechanics from Cornell University.

Paul Weiss is the director of the California NanoSystems Institute, professor of chemistry and biochemistry at UCLA, and Fred Kavli Chair in NanoSystems Sciences. His interdisciplinary research group includes chemists, physicists, biologists, materials scientists, electrical and mechanical engineers, and computer scientists. Their work focuses on the atomic-scale chemical, physical, optical, mechanical, and electronic properties of surfaces and supramolecular assemblies. They work to advance nanofabrication down to ever-smaller scales and greater chemical specificity to connect, operate, and test molecular devices.

Dr. Weiss has published extensively and holds numerous patents. He also has given more than 400 invited and plenary lectures. Before coming to UCLA, Dr. Weiss was a Distinguished Professor of Chemistry and Physics at the Pennsylvania State University. He received his S.B. and S.M. degrees in chemistry from MIT and his Ph.D. in chemistry from the University of California, Berkeley.

Joe Witte is a weather consultant to NASA on outreach to national and local television weather broadcasters. He also pursues graduate research studies at George Mason's Center for Climate Change Communication, where he is focused on engaging TV weathercasters as climate change educators. He is particularly interested in the visual rhetoric of climate change communication.

Mr. Witte is a fellow and councilor of the American Meteorological Society, and is a Fellow of the Explorers Club stemming from his field work in the arctic on Ice Island T-3. He began his career as a glaciologist for the U.S. Geological Survey, after which he worked at National Oceanic and Atmospheric Administration's Geophysical Fluid Dynamics Laboratory in Princeton, NJ. He was chief meteorologist at local TV stations in Seattle, Philadelphia, and New York City. He also acted as chief meteorologist and weather reporter at NBC where he worked at the TODAY Show and Nightly News. His weather field work included flights into three hurricanes. Mr. Witte earned an M.S. in atmospheric sciences from the University of Washington.

Gabrielle Wong-Parodi is a postdoctoral fellow at the Center for Climate and Energy Decision Making in the Department of Engineering and Public Policy at Carnegie Mellon University.

Dr. Wong-Parodi's area of expertise is in risk perceptions and communications. More specifically, she applies behavioral decision-making methods to address real-world energy and environmental problems to inform policy. Dr. Wong-Parodi has published papers on the risk perceptions of emerging technologies such as carbon capture and sequestration and smart grid technologies.

Previously, Dr. Wong-Parodi was a research associate with the Energy Efficiency Standards group at Lawrence Berkeley National Laboratory. Her most recent work at the lab was developing a model of the U.S. natural gas and coal system to be used to assess the economic and environmental impact of proposed climate change policies on federal residential appliance standards. Dr. Wong-Parodi received her B.A. in psychology and her M.A. and Ph.D. in energy and resources from the University of California, Berkeley.

Peter Zandan is the global vice chair and worldwide research practice group leader at Hill+Knowlton Strategies, a leading communications firm with 90 offices in 52 countries. He has been instrumental in developing the firm's reputation management and communications research offerings.

Prior to joining H+K Strategies, Dr. Zandan founded Intelliquest and served as its chairman and CEO for 15 years. Under his leadership, Intelliquest became the fastest growing market research company worldwide and was publicly traded on the Nasdaq. Dr. Zandan also founded and served as CEO of Zilliant, a successful venture-backed pricing optimization company. He has been named Ernst & Young's "Entrepreneur of the Year" and was selected by *Interactive Week* as one of the "Unsung Heroes of the Internet."

Dr. Zandan is a member of the Presidents' Circle of the National Academies and serves as a lifetime member of the McCombs School of Business Advisory Council. He earned his M.B.A. and Ph.D. in evaluation research from the University of Texas at Austin.

Kathleen M. Zelman is director of nutrition at WebMD, where she is responsible for overseeing diet, nutrition, and food information. In that role, Ms. Zelman is senior nutrition correspondent; writes features, columns, and diet book reviews; provides expert editorial review of diet and nutrition articles; and covers national meetings. Ms. Zelman also has extensive media experience, including cohosting a weekly radio program and serving as a national spokeswoman for the Academy of Nutrition and Dietetics (formerly the American Dietetic Association) for 12 years.

Ms. Zelman has received much recognition for her work. The Institute of Food Technologists awarded her the 2012 Media Award for Excellence in Consumer Journalism, and in 2011 she received the Nutrition Science Media Award from the American Society for Nutrition for fostering public understanding of nutrition issues. Ms. Zelman received a B.S. from Montclair State University and an M.P.H. from Tulane University.